T0122567

Advances in Intelligent Systems and Computing

Volume 849

Series editor

Janusz Kacprzyk, Polish Academy of Sciences, Warsaw, Poland
e-mail: kacprzyk@ibspan.waw.pl

The series "Advances in Intelligent Systems and Computing" contains publications on theory, applications, and design methods of Intelligent Systems and Intelligent Computing. Virtually all disciplines such as engineering, natural sciences, computer and information science, ICT, economics, business, e-commerce, environment, healthcare, life science are covered. The list of topics spans all the areas of modern intelligent systems and computing such as: computational intelligence, soft computing including neural networks, fuzzy systems, evolutionary computing and the fusion of these paradigms, social intelligence, ambient intelligence, computational neuroscience, artificial life, virtual worlds and society, cognitive science and systems, Perception and Vision, DNA and immune based systems, self-organizing and adaptive systems, e-Learning and teaching, human-centered and human-centric computing, recommender systems, intelligent control, robotics and mechatronics including human-machine teaming, knowledge-based paradigms, learning paradigms, machine ethics, intelligent data analysis, knowledge management, intelligent agents, intelligent decision making and support, intelligent network security, trust management, interactive entertainment, Web intelligence and multimedia.

The publications within "Advances in Intelligent Systems and Computing" are primarily proceedings of important conferences, symposia and congresses. They cover significant recent developments in the field, both of a foundational and applicable character. An important characteristic feature of the series is the short publication time and world-wide distribution. This permits a rapid and broad dissemination of research results.

More information about this series at http://www.springer.com/series/11156

Wai Keung Wong
Editor

Artificial Intelligence on Fashion and Textiles

Proceedings of the Artificial Intelligence on Fashion and Textiles (AIFT) Conference 2018, Hong Kong, July 3–6, 2018

 Springer

Editor
Wai Keung Wong
Institute of Textiles and Clothing
The Hong Kong Polytechnic University
Hunghom, Hong Kong

ISSN 2194-5357 ISSN 2194-5365 (electronic)
Advances in Intelligent Systems and Computing
ISBN 978-3-319-99694-3 ISBN 978-3-319-99695-0 (eBook)
https://doi.org/10.1007/978-3-319-99695-0

Library of Congress Control Number: 2018952621

This Springer imprint is published by the registered company Springer Nature Switzerland AG
The registered company address is: Gewerbestrasse 11, 6330 Cham, Switzerland

Contents

A Clothing Recommendation System Based on Expert Knowledge

Tao Yang, Jiao Feng, Jie Chen, Chunyan Dong, Youqun Shi and Ran Tao

Abstract Through summarizing expert experience and knowledge of clothing, the clothing recommendation system is developed based on a kind of clothing recommendation method. According to color matching rules, this method has refined the six factors that affect the customer's choice of clothing, establish the clothing knowledge base and clarify the recommendation rules. Considering the characteristics of the customers and the selection criteria, this system can make personalized clothing recommendation scheme for customers and ensure the rationality of the recommendation results.

Keywords Expert system · Cloth recommendation · Knowledge base

1 Introduction

With the increase of individualized wearable consciousness, clothing is not only the basic living demand, but also the important carrier to enhance self-taste and image.

T. Yang (✉) · J. Feng · J. Chen · C. Dong · Y. Shi · R. Tao
Donghua University, Shanghai 200051, China
e-mail: yangtao@dhu.edu.cn

J. Feng
e-mail: shirleyjiao@outlook.com

J. Chen
e-mail: heroxiaowanzi@qq.com

C. Dong
e-mail: 1210017759@qq.com

Y. Shi
e-mail: yqshi@dhu.edu.cn

R. Tao
e-mail: taoran@dhu.edu.cn

© Springer Nature Switzerland AG 2019
W. K. Wong (ed.), *Artificial Intelligence on Fashion and Textiles*,
Advances in Intelligent Systems and Computing 849,
https://doi.org/10.1007/978-3-319-99695-0_1

In the daily shopping, the various clothes on the e-commerce platform often make customers regret consumption [1].

Based on the above background, a personalized clothing recommendation system based on expert system has been designed and developed in this paper, which aims to guide consumers to choose the suitable clothing on the perspective of professional match. The main work is the application of clothing recommendation and match experience provided by experts in personalized cloth recommendation. We have refined six factors affecting customer dress, transformed the thinking mode of experts into the electronic knowledge base and recommendation process that computer can handle.

2   Clothing Recommendation Knowledge Base

The knowledge base in the clothing recommendation system mainly refers to the set of rules used by the system runtime, including the data information corresponding to the rules and the storage mode that rules can be transformed and processed by computers after summarizing experts' experience [2].

2.1   *Customer Characteristics and Clothing Elements*

Recommendation based on expert rules uses experts' knowledge which can map customers' needs to product features and take customers' attributes as the main consideration [3]. The recommendation system depends on two parts of customers and clothing, as shown in Fig. 1. Integrating with experts' years of experience to identify customer characteristics and clothing elements is the basis of knowledge base establishment and clothing recommendation realization.

Customer Characteristics Extraction. When experts design customer's image, the first consideration is the customer's skin color which influences the customer's suitable color range. The second point is to consider the body type. The correct use of clothing version can make up body defects. In order to clarify the customer's preferences, the style factor is considered. Each style has a corresponding theme color. Finally, the recommended results are given based on customer demand for clothing categories and specific colors [4].

The customers' information we collect are divided into two parts: objective factors and subjective factors, as shown in Fig. 1. Objective factors refer to the basic attributes of customers, including age, height, skin color, and measurements of chest, waist, and hips. Height and weight measurements are used to determine the customer's body type. Subjective elements are updated as the preferences of customer change, including style, preference color and category. Based on the above-mentioned rules, customer factors can be the following four options:

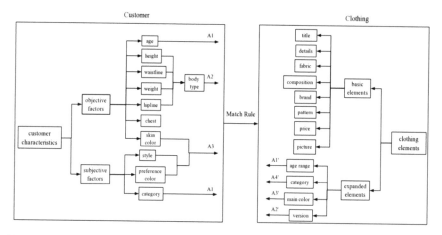

Fig. 1 Customer characteristics and clothing elements

- Age range. In addition to a small number of basic styles of clothing, clothing has its age range which is suitable for customers, the age of the customer determines the choice of clothing A4.
- Version range. Clothing version can help customers highlight the advantages of figure while concealing defects, so obtaining customer size data means that system can choose the appropriate range of clothing version A2.
- Color range. In customer characteristics, the three factors including skin color, style, and preference color are related to color. Skin color influences the color range of the customer. Style determines the theme color. Preference color defines color selection range. The combination of the three factors can generate the customer clothing's best color range A3.
- Category range. Customer's demand for a particular category is the most intuitive clothing conditions which can obtain the category range A1.

Clothing Elements Extraction. Clothing description is inseparable from the clothing name, fabric, composition, brand and other information, shown in Fig. 1, we consider this information as the basic elements. When the system recommends the clothing, the basic information is far from enough. In order to be consistent with customer factors, we join the four extended elements which contain category, age range, the main color, and clothing version. Expanded elements need to meet the system-specific classification and data requirements [5]. As an example, the primary classification of clothing is the cloth of upper body. The second classification for this primary classification is T-shirt, shirt, sweater and so on. Main color refers to the largest proportion of fabric color. The clothing version refers to the outer contour of the clothing. When matching the clothing for the customer, the customer's range of choice for clothing and clothing elements are be consistent one-to-one [6]. The clothing's category elements A1′ correspond to the customer's category range. The clothing's suitable age range elements A4′ corresponds to the customer's age range A4. The clothing's main color elements A3′ corresponds to the customer's color range A3. The clothing's version elements A2′ corresponds to the customer's version range A2.

2.2 Knowledge Base

Clothing knowledge base consists of three parts, including element library, color library, and recommendation rules algorithm, as shown in Fig. 2. The element library is divided into customer elements and clothing elements which store the classification of each factor and the corresponding color range. Color library includes the correspondence of Pantone and RGB, color similarity calculation method, and basic color expansion. The rule algorithm includes the clothing's main color identification algorithm and clothing matching rules. The matching rules are used to determine the matching between the customer and the clothing。The recommendation results are sorted according to the matching degree.

3 Clothing Recommendation

After the knowledge bases are decided, the system needs to search the clothing database for the clothing which meets the requirements according to the customer's physical characteristics.

3.1 Matching Degree Calculation

The matching degree quantifies the suitability of the garment under the user factor through a specific score. The high total matching degree indicates that the garment is more suitable for the user and will be recommended preferentially. The single clothing matching degree set C = {C1, C2, C3, C4} corresponds to the matching degree of style, skin color, body type, and age respectively, within the range of [0,1]. The total matching degree of clothing formula is as follows [7]:

$$C = C1 + C2 + C3 + C4 \tag{1}$$

For example, the age matching degree is used to determine the match between a user of a certain age and a suit for a certain age. The age matching degree is divided into four levels, as shown in Table 1. Taking a 26-year-old user as an example, the cloth with an age range of 25–29 is the best choice which matching degree is 1; the cloth with an age range of 18–24 or 30–34 is the second choice which matching

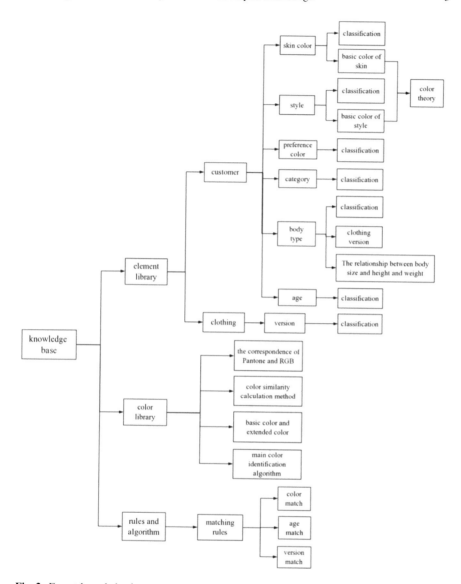

Fig. 2 Expert knowledge base

degree is 0.6 and so on, the cloth which age range is older than 40 is not recommended, and the match degree is zero.

Table 1 Age matching degree division (26-year-old user)

Age range that the cloth suits	Matching degree
25–29	1
18–24 or 30–34	0.6
35–39	0.3
≤7 or >40	0

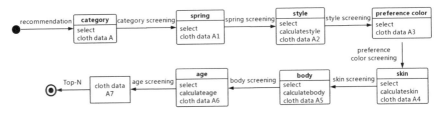

Fig. 3 Clothing filter process

3.2 Clothing Recommendation Process

Figure 3 shows the clothing filter process. A represents the system's clothing database. First, according to the filter of category factors, clothing data A1 is achieved. A1 excludes clothing which does not meet the category. On the basis of A1, the style is selected. The matching degree C1 of the clothing for the style is calculated. The clothing which is unsuitable to this style is excluded. The clothing data A2 is obtained. After preference color, skin color, body type, and age factors are selected, skin color matching degree C2, body shape matching degree C3 and age matching degree C4 are calculated. Data with a matching degree of zero is excluded. The final clothing data A7 is obtained which need to be sorted by the sum of C1, C2, C3, C4. The Top-N clothing can be recommended based on A7.

3.3 Clothing Recommendation Result

Taking into account the current user's usage habits, the system is developed based on mobile platform. For example, user A has a warm skin and an H-shaped figure. She is 23 years old and prefers vintage style. In the spring she needs a shirt with blue and purple color. The results are shown in Fig. 4. The recommended clothing needs to meet the direct filter conditions of the shirt category and blue-violet preference color. The loose clothes can make up for the shortcomings of her thin upper body. The main colors of both garments conform to the user style and the basic color of the skin color.

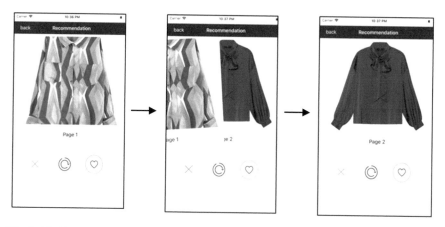

Fig. 4 Clothing recommend result in mobile

4 Conclusion

This article puts forward a clothing recommendation system based on expert rules for the problem of difficult to buy clothing. Compared with the traditional clothing recommendation system, the recommendation system based on expert rules can provide well-directed recommendation services to guide customers to purchase suitable clothing. Combined with the knowledge base, personalized recommendation results will be obtained after multiple factors screening.

Acknowledgments This work is sponsored by Dongguan City professional town innovation service platform construction project "Dongguan City Humen garment Collaborative Innovation Center" and Guangdong Province collaborative innovation and platform Environmental Science build of special funds NO. 2014B090908004.

References

1. Alfian, A.: The development framework of expert system application on indonesian governmental accounting system. In: International Conference on Computer Science and Artificial Intelligence, pp. 60–64 (2017)
2. Zhong, X., Liu, Z., Ding, P.: Construction and application of knowledge base based on hybrid reasoning. J. Comput. **35**, 761–766 (2012)
3. Cai, Z.: Advanced expert system: principles design and applications. Beijing Science Press, Beijing (2005)
4. Ying, Ni.: Clothing Image Design. China Textile Press (2012)
5. Salim, N.: Recommendation systems: a review. Int. J. Comput. Eng. Res. (2013)
6. Tewari, A.S.: Sequencing of items in personalized recommendations using multiple recommendation techniques. Expert Syst. Appl. **97**, 70–82 (2018)
7. Wagner, W.P.: Trends in expert system development: a longitudinal content analysis of over thirty years of expert system case studies. Expert Syst. Appl. **76**, 85–96 (2017)

Coordinated Optimization of Production and Delivery Operations in Apparel Supply Chains Using a Hybrid Intelligent Algorithm

Zhaoxia Guo, Jingjie Chen, Guangxin Ou and Haitao Liu

Abstract This paper addresses a coordinated optimization problem of production and delivery operations in apparel supply chains. A fleet of heterogeneous vehicles are used to deliver the accessories produced on parallel machines to a number of apparel production plants. We consider the flexible vehicle departure time between the production and distribution. A novel hybrid intelligent solution framework is proposed to solve this problem, by decomposition the optimum-seeking process is simplified and the computational complexity is reduced. The effectiveness of proposed framework is evaluated by numerical experiments. Experimental results show that the proposed solution framework exhibits better optimization performance in terms of the solution quality and computational time than other state-of-the-art algorithms.

Keywords Apparel supply chain · Production scheduling · Vehicle routing Intelligent algorithm

1 Introduction

In time-intensive apparel supply chains, garment accessory suppliers, such as embroidery plants and printing plants, need to process the embroidery or printing orders, and then deliver the finished orders to different local customers (garment plants) according to the customers' requirements. In order to reduce operating costs and enhance customer service levels, the production scheduling and delivery operations should be coordinated. Motivated by these practical applications, this study investigates an integrated optimization problem of production and distribution operations, called as the integrated production scheduling and vehicle routing with time windows (IPS-VR).

Z. Guo · J. Chen · G. Ou · H. Liu (✉)
Business School, Sichuan University,
Chengdu 610065, People's Republic of China
e-mail: haitaoliuch@gmail.com

© Springer Nature Switzerland AG 2019
W. K. Wong (ed.), *Artificial Intelligence on Fashion and Textiles*,
Advances in Intelligent Systems and Computing 849,
https://doi.org/10.1007/978-3-319-99695-0_2

9

In recent years, the IPS-VR problems have gained more and more researchers' attention [1–4]. Previous studies usually assume that orders need to be delivered immediately or at the fixed departure time after orders are finished [5–7]. These assumptions could lead to penalty costs of violating time windows of customers in practice. Compared to the immediate or fixed vehicle departure time, the flexible vehicle departure time can meet effectively the requirements of customer time windows because of the flexibility of vehicle departure time [7]. However, research on IPS-VR problems with flexible departure times in parallel machines has not been reported so far, although these problems exist widely in apparel supply chains. This study thus aims to investigate an IPS-VR problem with flexible vehicle departure times in apparel supply chains.

2 Problem Description

In an apparel supply chain environment, the supplier (accessory producer) receives various orders (indexed by $i \in \{1, 2, \ldots, I\}$, alias j) from its local customers (garment producer, indexed by i as well since each garment producer places one order only). These orders are processed on identical parallel machines (indexed by $m \in \{1, 2, \ldots, M\}$). The finished orders are delivered to local customers by a fleet of vehicles (indexed by $v \in \{1, 2, \ldots, V\}$) according to the given time windows of all customers. Different vehicles could have different loading capacities. We use q_v to denote the loading capacity of vehicle v. If the actual delivery time D_i of order i is less than the lower bound l_i of its time window, an earliness penalty cost will incur and we use E_i to denote the amount of early delivery time. If the actual delivery time D_i is greater than the upper bound u_i, a tardiness penalty cost will incur and we use T_i to denote the amount of late delivery time. Let c denote the transport cost per unit time, e and t denote the unit earliness and tardiness penalty cost for one order respectively. The investigated IPS-VR problem aims to determine the values of five decision variables so that the total supply chain cost is minimized. The cost is the summation of direct transport costs and the earliness and tardiness penalty costs of all orders. The five decision variables include: o_{mi} (it is 1 if order i is processed on machine m; otherwise it is 0); x_{ijm} (it is 1 if order j is processed immediately after order i on machine m; otherwise it is 0); y_{vi} (it is 1 if order i is transported by vehicle v; otherwise it is 0); z_{ijv} (it is 1 if customer j is visited immediately after visiting customer i by vehicle v; otherwise it is 0); and d_v (it denotes the flexible vehicle departure time of the vehicle v). The objective of this problem is to minimize the total supply chain cost.

$$\min F(o_{mi}, x_{ijm}, y_{vi}, z_{ijv}, d_v) = c \cdot \sum_{i=1}^{I} \sum_{j=1}^{I} \sum_{v=1}^{V} \left(r_{ij} \cdot z_{ijv} \right) + e \cdot \sum_{i=1}^{I} E_i + t \cdot \sum_{i=1}^{I} T_i$$

$$(1)$$

3 Hybrid Intelligent Optimization Framework

The investigated IPS-VR problem needs to determine the values of five decision variables: o_{mi}, x_{ijm}, y_{vi}, z_{ijv} and d_v. This problem is actually a multi-level optimization problem, which can be tackled by solving two sub-problems in an integrated and nested manner: a parallel machine scheduling sub-problem for determining the values of variables o_{mi} and x_{ijm}, and a distribution scheduling sub-problem for determining the values of variables y_{vi}, z_{ijv} and d_v. Both the sub-problems are intractable when problem sizes are large, because both of them are NP-hard.

This study tackles the investigated problem by decomposing this IPS-VR problem into sub-problems with smaller problem sizes. The orders transported by each vehicle need to be produced in turn in the production plant. Therefore, we can solve a vehicle assignment sub-problem first, which determines the values $\{y_{vi}\}$ of order assignment to vehicles. It is equivalent to the general assignment problem. Then we only need to handle parallel machine scheduling sub-problem that determines the values of variables o_{mi} and x_{ijm}, which is much easier to solve than the original parallel machine scheduling sub-problem because less orders need to be considered in each sub-problem. Next, we solve a new distribution scheduling sub-problem, which determines the values of variables z_{ijv} and d_v. By so doing, the investigated IPS-VR problem is decomposed into three simpler sub-problems with smaller problem sizes.

This study proposes a hybrid intelligent optimization (HIO) framework to solve the three sub-problems in a coordinated and nested manner by combining intelligent optimization techniques with heuristic procedures. This framework consists of an outer level optimization process and an inner-level optimization process. The outer-level optimization process aims to seek the best vehicle assignment $\{y_{vi}\}$ by using an intelligent optimization techniques, while the inner-level optimization process aims to seek the best values of other four variables o_{mi}, x_{ijm}, z_{ijv} and d_v. The flowchart of HIO framework is shown in Fig. 1. The first step is to initialize algorithm parameters used in HIO framework. These parameters include population size and other parameters used in the framework, such as crossover and (or) mutation probabilities in evolutionary algorithms. Step 2 generates the initial population by randomly assigning all orders to vehicles. Each individual in the population represents a vehicle assignment solution $\{y_{vi}\}$, which is a sequence of orders to be transported by vehicle v. Step 3 is to evaluate the performance of each individual in the initial population based on the inner level optimization process. Since each individual only determines decision variable y_{vi}, the corresponding other decision variables (o_{mi}, x_{ijm}, z_{ijv} and d_v) are to be determined by sub-steps shown in Fig. 1b. Step 3a is to determine the production assignment $\{o_{mi}\}$ and the processing sequences $\{x_{ijm}\}$ of orders. Then the makespans of orders delivered by each vehicle are calculated, after which it is to handle the distribution scheduling by determining the vehicle departure time d_v and routes z_{ijv} of vehicle v. The procedures involved in step 3 are skipped due to page limit. Steps 4–7 constitute the iterative process of HIO framework. Each iteration denotes a new generation of the outer-level optimization process. The new child individual is generated and evaluated respectively in steps 4 and 5. The procedure

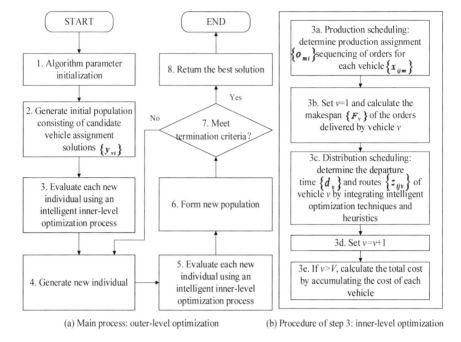

(a) Main process: outer-level optimization   (b) Procedure of step 3: inner-level optimization

Fig. 1 Hybrid multi-level intelligent optimization framework

in step 6 is the same as the procedure in step 3. Step 6 forms the new population
based on the fitness of parent and offspring individuals. The termination criterion
is checked in step 7 in each iteration. If the specified maximum iterations g_{max} is
reached, the iterative process of the HIO framework is terminated; otherwise, the
process returns to step 4 and continues to generate the new individual. Step 8 returns
the best solution individual in the current population as the best solution found by
the proposed HIO framework.

The HIO framework is a general solution framework for the investigated IPS-VR
problem. Under the HIO framework, various intelligent optimization techniques [8],
such as genetic algorithm (GA), tabu search (TS), evolution strategy and memetic
algorithm, can be used to seek the best solutions. The traditional GA [9] is used
as the outer level optimization process in this paper. Under the HIO framework,
the objective of parallel machine scheduling sub-problem is set to minimize the
makespan of all orders, which does not affect the final solutions to the investigated
IPS-VR problem. The BFD-LPT heuristic developed by Xu et al. [10] is adopted
to obtain the values of o_{mi} and x_{ijm}, because it can handle effectively the parallel
machine scheduling problem with the minimal makespan objective. We combine an
intelligent optimization technique with heuristic rules to obtain the values of z_{ijv}
and d_v, since seeking the optimal values of variable z_{ijv} is equivalent to solving a
traveling salesman problems with the objective of minimizing total transport cost

and penalty cost of violating time windows. This study adopts the TS, proposed by Fiechter [11], to obtain the best values of variables z_{ijv}.

4 Experiments and Comparison

To evaluate the effectiveness of the proposed HIO framework for investigated IPS-VR problem, a series of numerical experiments have been performed. Due to page limit, this section highlights one typical experiment only, which consider 35 customers, four vehicles and two machines.

The experiment data are generated based on the benchmark instances of the vehicle routing problem with time windows presented by Solomon [12].The loading capacity q_v of each vehicle is set to 100. For simplicity, the transport time among customers equals the corresponding Euclidean distance. The coordinate of the plant is (40, 50).

The penalty parameters e and t are equal to 1 and 3 respectively. The transport cost of per unit time c is equal to 2. The parameters including crossover probability, mutation probability, population size and maximal iteration number in GA are set to 0.8, 0.4, 150, and 100 respectively. The parameters including the tabu tenure, the length of tabu list, the number of candidate solutions and maximal iteration number in the TS are set to 10, 10, 20, and 80 respectively.

Table 1 shows that the best solutions and the corresponding results generated by the proposed HIO framework. The second column shows the makespan of the orders delivered by each vehicle. The third column shows the flexible vehicle departure times of each vehicle. The fourth column shows the route of each vehicle, in which "0" denotes the plant and other numbers denote the corresponding customer or order number. The fifth column shows each route's costs generated, which are the summation of the earliness and tardiness (E/T) penalty cost and the transport cost of each vehicle. The sixth column shows the total costs which are equal to the objective function value. It can be seen that some vehicles (e.g., vehicle 3) depart from the plant immediately after the orders are finished, while the departure times of most vehicles are rescheduled and greater than the makespan of their orders. It indicates the necessity of setting the vehicle departure time flexibly.

To evaluate the optimization performance of the proposed HIO framework, this study compared the performances of the proposed HIO framework with the genetic algorithm-based approach (called as UGA in this study) proposed by Ullrich [1]. The experimental setting is similar to that set by Ullrich [1], and the only difference is that the vehicle departure time is set to the makespan of orders transported by this vehicle in the UGA. The parameters in the UGA are set according to the recommendations from Ullrich [1]. The HIO framework reduced the total costs by 23.34% compared with the UGA. With the increase in problem sizes, the proposed HIO framework shows a higher superiority over the UGA.

Table 1 The best solution generated by the proposed HIO framework

Vehicle no.	Makespan	Departure time	Vehicle routes	Cost of each route	Total cost
1	140	156.07	$0 \to 9 \to 11 \to 15 \to 16 \to 17 \to 14 \to 12 \to 7 \to 6 \to 4 \to 5 \to 2 \to 0$	294.24	1420.39
2	240	245.53	$0 \to 24 \to 19 \to 23 \to 21 \to 18 \to 30 \to 29 \to 31 \to 33 \to 28 \to 27 \to 0$	401.21	
3	280	685.08	$0 \to 35 \to 26 \to 32 \to 34 \to 0$	349.48	
4	360	360	$0 \to 20 \to 22 \to 25 \to 13 \to 10 \to 8 \to 3 \to 1 \to 0$	375.45	

5 Conclusion

This paper addressed a coordinated optimization problem of production and delivery operations in apparel supply chains. A hybrid intelligent optimization framework was proposed to solve the investigated problem. Experimental comparisons were performed to validate the effectiveness of the proposed HIO framework. The results showed that the HIO framework was able to tackle the IPS-VR problem effectively by providing the better solutions than the Ullrich's approach [1]. Different optimization techniques could be embedded in this framework to construct other algorithms with better optimization performances for IPS-VR problems in future.

References

1. Ullrich, C.A.: Integrated machine scheduling and vehicle routing with time windows. Eur. J. Oper. Res. **227**, 152–165 (2013). https://doi.org/10.1016/j.ejor.2012.11.049
2. Pundoor, G., Chen, Z.L.: Scheduling a production–distribution system to optimize the tradeoff between delivery tardiness and distribution cost. Naval Res. Logist. **52**, 571–589 (2005). https://doi.org/10.1002/nav.20100
3. Moons, S., Ramaekers, K., An, C., Arda, Y.: Integrating production scheduling and vehicle routing decisions at the operational decision level: a review and discussion. Comput. Ind. Eng. **104**, 224–245 (2017). https://doi.org/10.1016/j.cie.2016.12.010
4. Chen, Z.-L.: Integrated production and outbound distribution scheduling: Review and extensions. Oper. Res. **58**, 130–148 (2010). https://doi.org/10.1287/opre.1080.0688
5. Gao, S., Qi, L., Lei, L.: Integrated batch production and distribution scheduling with limited vehicle capacity. Int. J. Prod. Econ. **160**, 13–25 (2015). https://doi.org/10.1016/j.ijpe.2014.08.017
6. Amorim, P., Guenther, H.O., Almada-Lobo, B.: Multi-objective integrated production and distribution planning of perishable products. Int. J. Prod. Econ. **138**, 89–101 (2012). https://doi.org/10.1016/j.ijpe.2012.03.005
7. Agnetis, A., Aloulou, M.A., Fu, L.-L.: Coordination of production and interstage batch delivery with outsourced distribution. Eur. J. Oper. Res. **238**, 130–142 (2014). https://doi.org/10.1016/j.ejor.2014.03.039
8. Eiben, A.E., Smith, J.: From evolutionary computation to the evolution of things. Nature **521**, 476–482 (2015). https://doi.org/10.1038/nature14544
9. Goldberg, D.E.: Genetic Algorithms in Search, Optimization and Machine Learning. Addison-Wesley Pub. Co., Boston, MA (1989)
10. Xu, D., Sun, K., Li, H.: Parallel machine scheduling with almost periodic maintenance and non-preemptive jobs to minimize makespan. Comput. Oper. Res. **35**, 1344–1349 (2008). https://doi.org/10.1016/j.cor.2006.08.015
11. Fiechter, C.N.: A parallel tabu search algorithm for large traveling salesman problems. Discret. Appl. Math. **51**, 243–267 (1994). https://doi.org/10.1016/0166-218x(92)00033-i
12. VRPTW Benchmark Problems. http://w.cba.neu.edu/msolomon/problems.htm

Intelligent Cashmere/Wool Classification with Convolutional Neural Network

Fei Wang, Xiangyu Jin and Wei Luo

Abstract It is generally believed that there are subtle differences in textures and diameters, between cashmere and wool fibers. Thus, automatically classifying the cashmere/wool fiber images remains a major challenge to the textile industry. In this proposal, we introduced a method that uses Convolutional Neural Networks (CNNs) to identify the two kinds of animal fibers. Specifically, a typical CNN was used to extract image features at first step. Then a region proposal strategy (RPS) was used to localize the fine-grained features from the images. We fine-tuned the CNN model by using the features selected by RPS. Experiments on cashmere/wool image set compared to different models verified the effectiveness of the proposed method for feature extraction.

Keywords Cashmere/wool · Subtle differences · Classification
Convolutional neural networks

1 Introduction

Historically, the identification of cashmere and wool has been a major issue for consumers and textile manufacturers. Cashmere, is a kind of luxury animal fibers [1] because they are difficult to obtain large quantities. Owing to the fact of the fibers' softness, luster, and scarcity, cashmere is one of the finest and popular animal hair fiber among the above-mentioned fibers [2, 3]. However, fine-descaled wool or stretched wool is used to adulterate cashmere [4]. Thus, the research for classifying

F. Wang · X. Jin (✉)
Donghua University, Shanghai 201620, China
e-mail: jinxy@dhu.edu.cn

W. Luo
South China Agricultural University, Guangzhou 510642, China

W. Luo
Key Laboratory of Intelligent Perception and Systems for High-Dimensional Information of Ministry of Education, Nanjing University of Science and Technology, Nanjing 210094, China

© Springer Nature Switzerland AG 2019
W. K. Wong (ed.), *Artificial Intelligence on Fashion and Textiles*,
Advances in Intelligent Systems and Computing 849,
https://doi.org/10.1007/978-3-319-99695-0_3

17

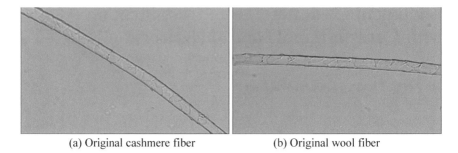

|(a) Original cashmere fiber|(b) Original wool fiber|

Fig. 1 Original microscope images of cashmere and wool fiber

cashmere and wool under consideration of model design has fundamental theoretical and practical meaning.

The high-quality cashmere fiber is ordinarily white, 35–50 mm in length, with a mean diameter of 15–19 mm. By contrast, Merino sheep's wool is usually 50–90 mm in length, with a mean fiber diameter of 18–25 mm [5]. Figure 1 shows raw images of cashmere and wool.

We can find that wool has a deeper texture than cashmere, and the two fibers have subtle different cross-sectional diameters.

However, the interested parties use phosphorus removal or stretch wool as a cashmere adulterant; and the diameter and the length of the cashmere fibers are significantly changed [6].

Traditionally, cashmere/wool fiber identification is carried out by experts on high-power Optical microscopy (OM) or scanning electron microscopy (SEM) [7, 8]. OM and SEM methods are considered as the most practical methods due to its high efficiency, and they cost little as well. The other type methods that receive a large audience and become practicable methods are DNA techniques [1, 9]. The DNA analysis method is reliable and objective as well as it can achieve mass testing. However, the whole process will cost one day, and the entire settings and instruments are relatively expensive compared to optical microscopy.

Various methods aim to classify cashmere/wool are based on Computer Vision (CV) technique [4, 10–12] they obtain the highest accuracy of 94.6%. Until recently, we have witnessed the great success of CNNs in fine-grained image classification, [13–18] (FIC) which refers to the task of classifying objects that belong to the same basic-level category. Alex Krizhevsky et al. [19] in the ImageNet LSVRC-2010 contest, trained a large, deep convolutional neural network to classify the 1.2 million high-resolution images into the 1000 different classes. They achieve a performance of top-1 and top-5 error rates of 37.5 and 17.0%.

Inspired by the literature above, this paper contributes to the field of cashmere/wool identification by developing a model of the CNN which extracts the features automatically and introduces a region proposal strategy (RPS) to predict object bounds and make scores at each position of the image.

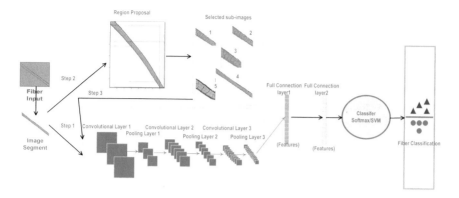

Fig. 2 The architecture of proposed CNN model

2 Method

The wool fibers and the cashmere fibers were supported by Ordos Group. The experimental images were acquired by using an optical measuring instrument. The optical microscopy system (CU-5) took photographs of the fibers at 10×50 magnifications. The dataset consisted of 2938 full-size fiber images (cashmere and wool), including 1705 cashmere images and 2082 wool images. The labels of the fiber have been given by seasoned experts of Ordos Group as well Method.

2.1 Overall Method

In our method, the cashmere/wool identification was carried out in three steps. First, fiber images were segmented to enhance features and remove noises. Second, we constructed a region proposal strategy which chose candidates from the sub-images that cut by the individual whole animal fibers. Finally, we employed sub-images selected by RPS as an enhancement input for the entire fiber images to the CNN classifier. Figure 2 presents the overview of the architecture of our CNN classifier.

Logistic regression function was chosen as the classifier to obtain the identification results. For the convenience of explanation, in the following sections, our model is abbreviated as CNN. The region proposal strategy which used for our CNN model is abbreviated as RPS.

2.2 Model Selected

Through the observation of the fiber sample images (as shown in Fig. 1a and b), it can be found that each fiber image contains a large area of invalid background fill. And the features extracted by CNN are calculated by pixels, thus the large area of invalid background pixels negatively affects the calculation and delivery of features. Region Proposal Method extracts the features of the fiber image from the four parameters of color, texture, size, and space overlapped. By this way, we can obtain the most effective feature expression regions in the above four aspects. In order to enhance the weights of the valid features, we construct a CNN method with Region Proposal Strategy (RPS). However, the model selected achieved a better performance than without using it.

2.3 Region Proposal Strategy

We use the selective search algorithm to obtain the most characteristic sub-image as an enhanced sample of fiber features. This section describes each step of the procedure in detail

1. Calculate the various color space of the image as the original candidate region (We consider (1) RGB, (2) gradation I, (3) Lab, (4) rgI (normalized rg channel plus gradation), (5) HSV, (6) rgb (normalized RGB), (7) C, (8) H (H channel of HSV) to calculate). The original candidate regions record as $R = \{r_1, r_2, \ldots, r_n\}$.
2. Initialize the similarity set $S = \emptyset$ (Here we use S to save the selected region proposal set).
3. Calculate a similarity between the two adjacent regions $R(r_i, r_j)$, and add it to the set S which stands for the similarity of the region R. The total similarities are calculated as shown in formula 1.

$$S(r_i, r_j) = a_1 S_{\text{color}}(r_i, r_j) + a_2 S_{\text{texture}}(r_i, r_j) + a_3 S_{\text{size}}(r_i, r_j) + a_4 S_{\text{fill}}(r_i, r_j) \quad (1)$$

$$a_i \in \{0, 1\}$$

In the formula 1, $S_{\text{color}}(r_i, r_j)$ here indicates Color Similarity (CS). $S_{\text{texture}}(r_i, r_j)$ indicates Texture Similarity (TS). $S_{\text{size}}(r_i, r_j)$ stands for Size Similarity (SS) and $S_{\text{fill}}(r_i, r_j)$ stands for Fit Similarity (FS).

4. The two regions r_i and r_j which has the highest similarity that found out from the similarity set S, are merged into one region r_t. In the meantime, the similarities originally calculated between adjacent regions r_i and r_j are removed from the similarity set S. Calculate the similarity between r_i and r_j its neighboring regions. The results are added to the similarity set S. At the same time, add the new region

Fig. 3 Region proposal
selected by selective search
algorithm

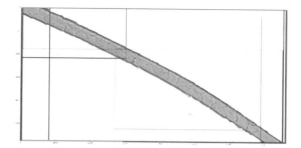

r_t.to the area set R. Iterate through the process above until the set S is removed
to $\emptyset$, which indicating that all the mergeable regions have been merged.

5. Obtain the location of each region r in set R, which is shown in Fig. 3. The five
   red bounding boxes are selected to represent the fiber's features.

Since the region proposal selected by the algorithm above is calculated from the
color, texture, size and the fit, it can be considered as the most distinctive features
of the entire fiber image. Thus, we take advantage of these features and training
the classifier by stretch the RPS-image (Selected by Region Proposal) to the same
dimensions as the original image and feed them to CNN.

3   Experiments and Results

After the knowledge base are decided, the system needs to search the clothing
database for the clothing which meets the requirements according to the customer's
physical characteristics.

3.1   Experimental Setup

The matching degree quantifies the suitability of the garment under the user factor
through a specific score. The high total matching degree indicates that the garment
is more suitable for the user and will be recommended preferentially. The single
clothing matching degree set $C = \{C1, C2, C3, C4\}$ corresponds to the matching
degree of A sample set contains 1980 fiber images (cashmere, Chines fine native
wool, and Merino wool) are prepared. In terms of classification problem, keras with a
theano (1.0) backend are utilized to build a model for classifying the cashmere/wool
images by Python language. The key parameters of the computer, calculating the
proposed model, are Intel (R) Xeon (R) E3-1231 v3 CPU@3.4 GHz and GPU of
Nvidia GTX960 with 4G Ram, and the operating system was Ubuntu 14.04 LTS
(64bit).

Table 1 Comparison between four CNN models

Methods	Training accuracy (%)	f-measure	Testing accuracy (%)	f-measure
Alex-Net	83 ± 5.5	0.79	78 ± 2.7	0.76
CNN (step 1 only)	84.2 ± 6.5	0.93	79 ± 3.7	0.93
CNN (step 2 only)	62.1 ± 11.1	1.13	55.5 ± 4.3	0.85
CNN(with RPS refine)	91.8 ± 3.5	0.52	91.7 ± 2.8	0.44

The first experiment analysis by using our CNN model and other model is executed separately. In addition, we also change the training and validation set's ration to verify the robustness of the proposed algorithm.

3.2 Performance Evaluation

3.2.1 Results on Different Models

To analyze the effect of the CNN architecture used to extract features at each step, we compare to Alex-Net [19]. Table 1 shown the experiment results in details. In all cases, using our proposed RPS could significantly improve the accuracy.

Table 1 represents the comparison of typical Alex-Net and use of our CNN structure (made use of step 1 or step 2 or entire process) separately over the five times tests. The results indicated that the individual use RPS or entire-image features could not get the best performance. Also, the traditional architecture needed structural modification to suitable for fiber texture sensitivity cashmere/wool classification. With the help of RPS extracted features of step 2, the accuracy achieves the best performance at 94.5%. The same situation is also observed in the performance of the f-measure. The accuracy and the f-measure have some fluctuations in each network and architecture, but the number of fluctuations within the interval of 5% which indicate the method's stability. However, comparing of using features extracted by step 1 or step 2 separately, our entire process employed the features of both, that can enhance each other's performance. The theory and experiments indicated that cashmere/wool contained the subtle difference in texture features. However, the texture is the good factor for the classification of cashmere/wool.

Table 2 The performance of CNN (entire process) in different blend ratio

Sample groups no	Blend cashmere/wool	Blend ratio (%)	Training accuracy (%)	Validating accuracy (%)
1	839/767	52.2/47.8	92.8±3.7	84.6
2	765/931	45.1/54.9	92±2.9	93.6
3	854/787	52.0/48.0	90.2±4.6	92.3
4	679/827	45.1/54.9	90.5±2.3	83.7
5	544/721	43.0/57.0	90.7±2.5	91.2
6	743/568	56.7/43.3	89.5±3.6	89
7	377/982	27.7/72.3	73.6±2.1	73
8	515/903	36.3/63.7	91.4±3.5	79.4
9	433/574	43.0/57.0	93.8±3.5	87.7
10	828/741	52.8/47.2	93.5±2.5	92.6

3.2.2 Results in Different Blend Ratios

Further, in order to evaluate the reliability of CNN, in experiment three, we employed ten groups of randomly selected sample sets to test and verify the results (Table 2).

As shown above, the training accuracies were almost all exceeded 90%. However, the validating accuracy, all exceed 75%. Moreover, all the fluctuations of the training accuracies were all below 3.7%. There were six times which the validating accuracy were below 90% on the condition that the samples of cashmere and wool were out of balance. According to the results above, we concluded that our CNN was a robust and stable network for cashmere/wool identification in different blend ratios. Hence, our CNN enhanced by RPS is a stable method for cashmere and wool classification strategy.

This article puts forward a clothing recommendation system based on expert rules for the problem of difficult to buy clothing. Compared with the traditional clothing recommendation system, the recommendation system based on expert rules can provide well-directed recommendation services to guide customers to purchase suitable clothing. Combined with the knowledge base, personalized recommendation results will be obtained after multiple factors screening.

3.3 Conclusion

In this paper, we construct a CNN method with Region Proposal Strategy (RPS) to classify wools/cashmere images. The method enhances the model's feature learning ability by augmenting fine-grained samples through the designed RPS module. Experiments indicate that the RPS significantly improved the performance of CNN extracted features in the cashmere/wool identification task and validate the robustness and effectiveness of our method. The evaluation of different configurations further signifies the samples and features in different scales, and levels could complement each other and collaborate to achieve better performance.

Acknowledgements This work was supported in part by the National Natural Science Foundation of China under Grant 61702197, in part by the Natural Science Foundation of Guangdong Province under Grant 2017A030310261, in part by the Key Laboratory of Intelligent Perception and Systems for High-Dimensional Information of Ministry of Education under Grant JYB201708.

References

1. Kerkhoff, K., Cescutti, G., Kruse, L., Müssig, J.: Development of a DNA-analytical method for the identification of animal hair fibers in textiles. Text. Res. J. **79**, 69–75 (2009). https://doi.org/10.1177/0040517508090488
2. Subramanian, S., Karthik, T., Vijayaraaghavan, N.N.: Single nucleotide polymorphism for animal fibre identification. J. Biotechnol. **116**, 153–158 (2005)
3. Guo, D., Li, F., Zhang, X., Quan, Z., Zhao, Y.: Study on breeding of new line of cashmere and meat for Liaoning Cashmere Goat. Mod. J. Anim. Husb. Vet. Med. (2017)
4. Zhong, Y., Lu, K., Tian, J., Zhu, H.: Wool/cashmere identification based on projection curves. Text. Res. J. **87**, 1730–1741 (2017). https://doi.org/10.1177/0040517516658516
5. Phan, K.H., Wortmann, F.J.: Cashmere definition and analysis: scientific and technical status. In: The 3rd International Symposium on Specialty Fibres; 2004; DWI Aachen, Germany (2004)
6. Ma, H.Y., Tian, Z.F., Hong, X., et al.: Analysis on cashmere quality and morphology changes in different places of origin. Rep. Erdos Group China (2015)
7. Yang, G.F., Fu, Y., Hong, X., et al.: Discussion on the SEM/ OM in cashmere identification. China Fiber Insp. 17–20 (2006)
8. Lan-Zhi, H.E., Chen, L.P., Wang, X.M.: Detection methods of distinguishing cashmere and wool fibers. Prog. Text. Sci. Technol. (2008)
9. Tang, M., Zhang, W., Zhou, H., Fei, J., Yang, J., Lu, W., Zhang, S., Ye, S., Wang, X.: A real-time PCR method for quantifying mixed cashmere and wool based on hair mitochondrial DNA. Text. Res. J. **84**, 1612–1621 (2014). https://doi.org/10.1177/0040517513494252
10. She, F.H., Kong, L.X., Nahavandi, S., Kouzani, A.Z.: Intelligent animal fiber classification with artificial neural networks. Text. Res. J. **72**, 594–600 (2002). https://doi.org/10.1177/004051750207200706
11. Shi, X., Yu, W.: A new classification method for animal fibers. In: Audio, Language and Image Processing, 2008. ICALIP 2008. International Conference on. pp. 206–210. IEEE (2008)
12. Cai-Xia, M.A., Liu, X.N., Liu, F.: A research on cashmere automatic identification method based on statistical analysis. Wool Text. J. (2014)
13. Branson, S., Van Horn, G., Belongie, S., Perona, P.: Bird species categorization using pose normalized deep convolutional nets. ArXiv Prepr. ArXiv14062952. (2014)
14. Zhang, N., Farrell, R., Iandola, F., Darrell, T.: Deformable part descriptors for fine-grained recognition and attribute prediction. Presented at the December (2013)
15. Ge, Z., McCool, C., Sanderson, C., Corke, P.: Modelling local deep convolutional neural network features to improve fine-grained image classification. In: Image Processing (ICIP), 2015 IEEE International Conference on. pp. 4112–4116. IEEE (2015)
16. Wah, C., Van Horn, G., Branson, S., Maji, S., Perona, P., Belongie, S.: Similarity comparisons for interactive fine-grained categorization. In: Proceedings of the IEEE Conference on Computer Vision and Pattern Recognition. pp. 859–866 (2014)
17. Ge, Z., McCool, C., Sanderson, C., Corke, P.: Subset feature learning for fine-grained category classification. In: Proceedings of the IEEE Conference on Computer Vision and Pattern Recognition Workshops. pp. 46–52 (2015)

18. Chai, Y., Lempitsky, V., Zisserman, A.: Symbiotic segmentation and part localization for fine-grained categorization. Presented at the December (2013)
19. Krizhevsky, A., Sutskever, I., Hinton, G.E.: ImageNet classification with deep convolutional neural networks. In: International Conference on Neural Information Processing Systems. pp. 1097–1105 (2012)

Yarn Quality Prediction for Spinning Production Using the Improved Apriori Algorithms

Xianhui Zeng and Pengcheng Xing

Abstract With the advent of the era of big data, intelligent manufacturing is becoming the developing direction of textile and clothing industry. Many textile enterprises are exploring the intelligent techniques to improve production efficiency and product quality. This paper focuses on yarn quality prediction based on the big data of spinning production. First, the association rules algorithms, Aproiori and I_Apriori algorithms, which are commonly used for intelligent prediction in spinning production, are analyzed. Then, aiming to overcome their disadvantages such as low efficiency, time consuming, big error of prediction results under big data, a global optimization strategy based on genetic algorithm is proposed. This strategy optimizes the global search process of Aprioir algorithm for pruning the frequent itemsets by introducing the genetic algorithm, which can avoid the local optimal solution in the search process. Finally, based on the big real production data collected from the spinning factory, the effectiveness of the proposed Apriori algorithm are investigated and compared with the normal Apriori algorithm. The result indicates that the improved algorithm has better efficiency and more accurate prediction results and is good at dealing with the big data environment.

Keywords Yarn quality prediction · Spinning production · Apriori algorithms
Genetic algorithms · Big data

X. Zeng (✉) · P. Xing
School of Information Science and Technology, Donghua University,
Shanghai, China
e-mail: xhzeng@mail.dhu.edu.cn

X. Zeng · P. Xing
Education Engineering Center of Digital Textile Technolgoy,
Shanghai, China

© Springer Nature Switzerland AG 2019
W. K. Wong (ed.), *Artificial Intelligence on Fashion and Textiles*,
Advances in Intelligent Systems and Computing 849,
https://doi.org/10.1007/978-3-319-99695-0_4

1 Introduction

Quality prediction, as one of the most advanced measure for quality control, is very important for textile production. Correct prediction can decrease cost significantly and improve the production efficiency. The traditional yarn prediction methods is based on the technical experiments by judging the feasibility of cotton assorting subjectively, which can result in the unsatisfied product, increased cost and quality fluctuation [1].

In recent years, increasingly importance has been attached to quality prediction of textile product by entrepreneurs [2]. A lot of algorithms, such as classifying, regress, and forecasting have been applied into the intelligent quality prediction system for textile production. For example, Sun proposes the neural network model to forecast the index of yarn quality [3].

With the deepening of the intelligent prediction control research, many excellent intelligent algorithms have been proposed and applied into textile industry. Among them, Apriori algorithms have been proved to be good at quality prediction for production decision [4]. However, under the circumstance of big data, Apriori algorithms still have some limitations, (1) a lot of candidate itemsets may be generated during scanning frequent itemsets, which may increase the computing time of the algorithm; (2) the database may be accessed many times while computing the support degree of the candidate itemsets. Due to tons of data in textile production, the efficiency of the algorithm may be decreased significantly and the I/O load of the system may become heavier; (3) it cannot ensure the complete mining from the data for searching association rules only by setting support degree and confidence degree, which may result in some strong association rules not agreeing with the real circumstance [5]. In summary, the Apriori algorithms cannot meet the requirement of the big data for the textile intelligent manufacturing.

Aiming to overcome the disadvantages of Apriori algorithms such as low efficiency, time-consuming, big error of prediction results under big data, this paper proposes a global optimization strategy based on genetic algorithm, which optimizes the global search process of Aprioir algorithm for pruning the frequent itemsets by introducing the genetic algorithm, which can avoid the local optimal solution in the search process. Using the big real production data collected from the spinning factory, the effectiveness of the proposed Apriori algorithm are investigated and compared with the normal Apriori algorithm. The experiment results show that the improved algorithm has better efficiency and more accurate prediction results and is good at dealing with the big data environment.

2 Apriori Algorithms and I_Apriori Algorithms

2.1 Apriori Algorithms

The principle of Apriori algorithms is to search set $k + 1$ based on set k based on prior knowledge of frequent itemset, until the end of the frequent itemsets [6] by a layer-by-layer searching way. Then, association rules are generated from frequent itemsets based on confidence thresholds [6]. The calculation of frequent itemset consists of two steps, connecting and pruning.

2.1.1 Connecting

The candidate frequent $(k + 1)$ itemset C_{k+1} is generated by connecting k-itemset L_k with itself. The connecting rule is as follows:

Whether any two subsets (l_a, l_b) in k-itemset L_k is connectable or not, it depends on one condition: if their former $(k - 1)$ items are the same, then they are connectable. That is,

$$(l_a[1] = l_b[1]) \wedge (l_a[2] = l_b[2]) \wedge \ldots \wedge (l_a[k - 1] = l_b[k - 1]) \tag{1}$$

So, the connecting result of l_a and l_b is

$$l_a[1]l_a[2] \ldots l_a[k - 1]l_a[k]l_b[k] \tag{2}$$

2.1.2 Pruning

Pruning consists of two steps. The first step is to calculate all the items of candidate k-itemset C_k to get their $(k - 1)$-itemsubset according to the property "all non-empty subsets of any frequent itemset are frequent". So, the frequent subsets are determined. The second step is to scan database to calculate support degree of every candidate item of k-itemset C_k. Then compare with threshold value of support degree and delete items less than the value to obtain the final frequent k-itemset L_k.

2.2 I_Apriori Algorithms

Apriori easily overlooks the negative correlation of rules. For example, one strong association rule $(A \Rightarrow B)$ fulfills the threshold value of support degree. But its confidence coefficient of negative association rule is also very big. The reason could be negatively correlated inhibitory effects or mutual independency between itemsets.

Thus, this rule is mutually contradictory and wrong. Based on the above disadvantage, I_Apriori algorithms are proposed based on the interestingness index model.

$$
\begin{aligned}
\text{interest} &= (A => B) = \text{Conf}(A => B) = \text{Conf}(\bar{A} - B)\\
&= \frac{P(AB)}{P(A)} - \frac{P(\bar{A}B)}{P(\bar{A})}\\
&= \frac{P(AB) - P(AB)P(A) - P(\bar{A}B)P(A)}{P(A)[1 - P(A)]}\\
&= \frac{P(AB) - P(A)P(B)}{P(A)[1 - P(A)]}
\end{aligned}
\tag{3}
$$

The range of interestingness is $[-1, 1]$. That means A has a positive effect on B when interest $(A \Rightarrow B) > 0$. Meanwhile, the closer to 1 the interestingness is, the stronger the correlation between A and B is. A has a negative effect on B when interest $(A \Rightarrow B) < 0$. So, the closer to -1 the interestingness is, the stronger the correlation between $\bar{A}$ and B is. The negatively correlated rule of rule $(A \Rightarrow B)$ (rule $\bar{A} \Rightarrow B$), which is not ignored, is mutually contradictory and could be deleted. A and B are independent when interest $(A \Rightarrow B) = 0$.

I_Apriori effectively removes bad strong association rules. However, like Apriori, when searching for frequent itemsets, the database still has to be traversed on a large scale, and it is also impossible to avoid the repeated combination of a large number of candidate subsets. This paper proposes a global optimization algorithm based on GA aiming at solving these drawbacks.

3   The Improved Apriori Global Optimization Algorithm Based on Genetic Algorithm

Compared with the traditional Apriori algorithm, the improved Apriori algorithm has been optimized in two aspects. The first is to adjust frequent itemsets before performing the pruning step to reduce the number of candidate itemsets in order to improve efficiency. The second is to search the frequent itemsets using the genetic algorithm which ensures global optimization.

3.1   Strategy of Pruning Frequent Itemsets

As it shows that the second property of Apriori algorithm is an element of an element that should be a frequent k-itemset whose occurrence in a frequent k-itemset must be no less than k times. Otherwise, the itemset containing this element cannot produce a candidate $(k + 1)$-itemset [7].

Table 1 Data interval code table of physical properties of cotton

Array element	A [1]	Mark	A [2]	Moisture regain/%
Item	I_1		I_2	
Value case	Value 1	2.00–2.20	Value 1	7.0–7.6
	Value 2	2.21–2.40	Value 2	7.7–8.2
	Value 3	2.41–2.60	Value 3	8.3–8.8
	Value 4	2.61–2.80	Value 4	8.9–9.4
	Value 5	2.81–3.00	Value 5	9.5–10.0
	0	None	0	None

After the frequent k-itemset is generated, the second property of the Apriori algorithm is used to simplify the number of itemsets to be used to perform the connection pruning step. Thereby, this contributes to reducing the number of repeated combinations in the process of connection generation of the candidate $(k+1)$-itemset and optimizing execution time. It is expected that the improved Apriori algorithm can reduce the number of scans by half [8].

3.2 Strategy of Global Optimization Based on Genetic Algorithm

The core task of the Apriori algorithm is to find all the frequent itemsets. The global search procedure can be optimized using genetic algorithm, which can greatly improve the efficiency of the Apriori algorithm. Next, the global optimization strategy is introduced through the actual problem of yarn quality prediction from textile production.

First, the data in textile production is coded. For example, the factors affecting the single strength of cotton yarn include mark, grade, moisture regain, and other physical attributes. As such data types collected by the equipment are non-Boolean data, and the value of each attribute is continuous and not fixed. So the attribute value needs to be divided into intervals according to historical data and actual conditions, and these intervals are defined as "value 1", "value 2", …, "value n". As is shown in Table 1.

The fitness function is used to evaluate the pros and cons of individuals and determine whether this individual can enter the next generation. Therefore, the definition of fitness function is the key to the algorithm. Whether a itemset is frequent or not depends on its support degree. Thus, support degree can be used to define fitness function.

In general, the fitness function is defined as,

Table 2 The interval and weight values for single-strength attribute

A[n]	Single-strength attribute	Weight
Value 1	12.0–12.7	4
Value 2	12.8–13.5	2
Value 3	13.6–14.3	1
Value 4	14.4–15.1	1
Value 5	15.2–15.9	2
Value 6	16.0–16.7	4

$$\text{Fit}(X) = \frac{\text{Supp}(X)}{\text{MinSupp}}, \qquad (4)$$

where, Supp(X) represents the support degree of the current itemset X, and MinSupp represents the minimum support degree threshold.

However, since the goal of the algorithm is to seek the single best relevance rule, we need a rule with a bigger single-strength attribute. At the same time, the association rule with a low single-strength attribute is also valuable. The reason is that these rules can be used to find what caused a low single-strength attribute and avoid it. Therefore, based on the formula (4), the fitness function is redefined by setting the weight of the single-strength attribute, as shown in formula (5).

$$\text{Fit}(X) = \frac{\text{Supp}(X_1, X_2, \ldots, X_{n-1}) + W_{1\sim 6}}{\text{MinSupp} + W_{1\sim 6}}, \qquad (5)$$

where, Supp($X_1, X_2, \ldots, X_{n-1}$) is the support degree of all the other attribute combinations except the single-strength attribute. $W_{1\sim 6}$ are the weights of the single-strength attribute, in which the single-strength attribute data is set as six intervals. The higher or lower the single-strength attribute has a greater weight. Attribute value in the middle has the lowest weight. Table 2 is the interval for single-strength attribute interval and its weight values.

According to the size of the fitness function, the genetic operation of selection, crossover and mutation are performed to generate the next-generation rule. After repeated iterations, a set of rules is obtained until the termination condition is satisfied. Finally, all the generated rules are performed screening and extraction using the confidence and interest degree. The operational flow is shown in Fig. 1.

4   Experimental Results and Discussion

In order to evaluate the performance of the improved Apriori algorithm, the real-life yarn quality prediction in spinning production was used for testing. First, the performance indicators of cotton fibers that affect the quality of yarn were determined [9]. Then the values of the index were collected and quantified. With these data,

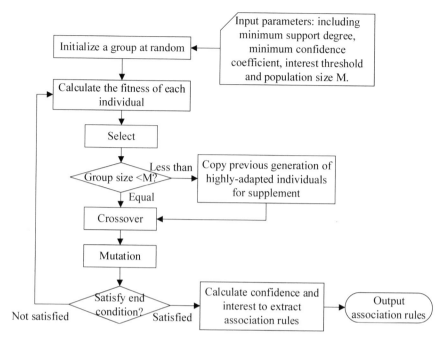

Fig. 1 Operational flow of the optimization algorithm

the algorithms were used to quantitatively predict the yarn quality according to the characteristics of the textile process. Three algorithms, Apriori algorithm, I_Apriori algorithm and global optimization algorithm based on genetic algorithm, was tested and simulated. The prediction performance of the three algorithms was analyzed and compared.

The experimental data were provided by the cotton spinning company Located in Zhejiang province. A total of 1500 pieces of data were used to form a single-strength training test sample of cotton spinning yarn. This group of data was obtained by 18.2 tex yarns of carding processes in the rotor spinning rapid spinning system. The extracted sample data are shown in Table 3.

The test samples are used to verify the prediction performance of the three algorithms. The experiment procedure follows the single variable principle. That is, under the same support and confidence conditions, the predicted values of the single-strength data are obtained through the three kinds of algorithms, and are recorded together with the actual values. Among them, the test samples for each group were randomly selected five data from the database (each piece of data consists of 12 items including marks, grade, moisture regain, etc.), and a total of 10 groups of tests were conducted.

The results of each test are plotted using scatter chart where the x axis represents the true value and the y axis represents the predicted value, as shown in Fig. 2. The

Table 3 The extracted sample data for the experiments

Mark	Grade	Moisture regain	Force	Main length	Uniform	Quality length	Micro-naire	Dirt per-centage	Maturity	Flock	Tenacity	Single strength
2.32	2.356	8.2	3.45	29.52	1102	32.62	4.61	2.5	1.54	10.66	20.23	12.6
2.01	2.562	8.9	3.78	29.36	1040	32.81	4.71	1.4	1.52	11.25	18.9	12.9
2.51	2.895	9	3.87	29.89	1134	33.01	4.59	1.9	1.62	11.89	19.56	13.6
2.01	2.452	7.1	3.56	30.38	1042	33.25	4.49	2.2	1.49	11.12	19.42	15.9
2.81	2.829	8.2	3.46	30.45	1013	32.45	4.56	2.4	1.6	12.23	19.23	16.5
2.46	2.459	8.6	3.25	29.56	1060	32.96	4.62	2.3	1.62	12.01	19.12	16.3
2.61	2.798	8.4	3.62	29.89	1134	31.95	4.66	2.3	1.52	12.62	18.23	14.7
2.32	2.789	9.6	3.68	29.05	1117	31.5	4.59	2.4	1.59	11.52	18.56	14.9
2.51	2.569	9.1	3.87	28.93	1076	31.45	4.62	2.1	1.57	11.45	18.95	15.4
2.22	2.989	8.1	3.6	30.54	1144	32.85	4.56	2.5	1.49	11.62	19.02	14.9
2.41	2.229	7.6	3.59	30.12	1029	30.12	4.59	2.5	1.62	11.89	20.01	16.1
2.36	2.829	8.4	3.34	29.89	988	30.89	4.69	2.5	1.52	10.89	19.45	13.6
2.58	2.346	9.1	3.56	29.31	1006	31.56	4.71	2.3	1.59	11.45	18.95	14.2
2.56	2.156	7.9	3.5	30.12	1027	32.45	4.65	2.2	1.62	11.62	19.45	15.3
2.61	2.874	8.1	3.72	29.58	1134	33.04	4.72	2.2	1.49	12.05	19.78	14.2
2.74	2.529	8.6	3.3	29.78	1086	31.85	4.62	2.1	1.56	12.42	18.45	15.2
2.52	2.529	7.1	3.34	29.44	1060	33.06	4.56	2.1	1.61	11.56	19.99	14.1
2.55	2.529	9	3.45	29.8	1061	32.68	4.59	2	1.62	11.78	20.21	14.6
2.52	2.365	8.8	3.65	30.38	1013	32.98	4.5	1.9	1.67	11.89	20.01	15.9
2.12	2.985	8.7	3.95	30.06	1032	32.45	4.43	2.4	1.71	11.2	19.25	16.2
2.83	2.456	8.2	3.12	30.22	1048	33.2	4.58	2.2	1.59	11.23	20.23	15.7
2.49	2.452	8.3	3.23	29.99	1077	33.12	4.44	2.3	1.54	10.89	19.45	12.9
2.41	2.874	7.8	3.55	29.41	1106	32.65	4.72	2.1	1.64	10.9	19.87	13.6
2.23	2.145	9.1	3.45	30.15	987	32.45	4.67	1.8	1.71	11.32	19.24	13.9

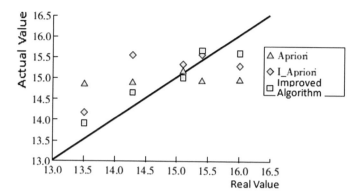

Fig. 2 The prediction results of three algorithms

slashes in the graph represent trajectories where the true and predicted values are absolutely equal.

The 10 groups of experimental results were analyzed by one-way ANOVA, which is used to analyze whether different algorithms had a significant influence on the prediction effect. The mean square error is calculated by formula (6).

$$\sigma = \sqrt{\frac{1}{5}\sum_{i=1}^{5}\left(x_{predict} - x_{true}\right)^2} \tag{6}$$

That is, the square root of the arithmetic mean of the difference between the predicted value and the true value, which can reflect the dispersion of the predicted value obtained by an algorithm from the true value, and is an important indicator of the accuracy of the prediction.

Through the test analysis of 10 groups of experiments, the final results are obtained. The mean square error of the Apriori algorithm is 3.131, the mean square error of the I_Apriori algorithm is 2.862, and the mean squared error of the optimization algorithm based on the genetic algorithm is 1.11. Therefore, it can be seen that the global optimization algorithm proposed in this paper is much better than the traditional Apriori algorithm and I_Apriori algorithm in accuracy. The traditional Apriori algorithm cannot meet the actual production needs. Its dispersion of predicted values is too large to predict the quality of spinning products. The I_ Apriori algorithm modifies the traditional algorithm to a certain degree, and the mean square deviation of the predicted value decreases, but the prediction effect is still unsatisfactory. Compared with the Apriori algorithm and the I_Apriori algorithm, the global optimization algorithm's prediction performance has been significantly improved, and the prediction result is more ideal. Therefore, the performance of the improved algorithm is far better than the traditional algorithm.

5   Conclusion

The paper proposes an improved Apriori algorithm with global optimization strategy, which prunes the frequent itemsets by introducing the genetic algorithm, which can avoid the local optimal solution in the search process. By comparing with the traditional Apriori algorithms and I_Apriori algorithms, the improved algorithm has better efficiency and more accurate prediction results and is good at dealing with the big data environment. The paper also applies the improved Apriori algorithm into yarn quality prediction for spinning production. The experiments show the satisfied results have been obtained. The improved algorithm can be used to process the big data by decreasing the computing time significantly and improving the prediction accuracy.

Refrences

1. Wu, X.: Research and Application on Quality Prediction Techniques of Textile Production. Donghua University, Shanghai (2009)
2. Wang, K.: Quality Prediction of Combed Wool Yarn Based on Computing Intelligence. Donghua University, Shanghai (2009)
3. Sun, H.: Yarn Quality Analysis and Prediction. Suzhou Univserity, SuZhou (2004)
4. Wang, D.: Optimization Research of Apriori Algorithms Based on Cloud Computing and Medical Big Data. Beijing University of Posts and Telecommunications, Beijing (2015)
5. Xu, Z., Liu, M., Zhang, S., et al.: Three optimization methods of Apriori algorithms. Comput. Eng. Appl. **6**(36), 190–193 (2004)
6. Chen, D.: The Correlation Analysis Research of Big Data Using Apriori Algorithm. China University of Geosciences, Beijing (2016)
7. Ouyang, T.: A kind of association rules improved algorithm based on genetic algorithm. J. Hangzhou Dianzi Univ. **9**(5), 79–81 (2015)
8. Xiao, D., Yang, L.: Association rules data mining based on genetic algorithm. Commun. Technol. **1**(43), 205–207 (2010)
9. Li, L.: Research and Application on Data Mining Technique for Lean Production. Donghua University, Shanghai (2010)

Everybody Immersive Fashion_Human–Computer Interaction in VR

Kate Kennedy

Abstract This paper discusses preliminary research into the use and application of virtual reality (VR) as an immersive design environment for conceptual and technical development of fashion and apparel, for individual bodies. The aim of the research was to investigate participant engagement and response to first, immersion in the VR environment and second to avatars created from 'real' 3d body scan data as an alternative body format to the average, standard or ideal 'fashion' body. The research was undertaken in VR workshops, *Everybody: Immersive Fashion Design*, at the *Slow Fashion Studio* during the RMIT Gallery/Goethe-Institut's *Fast Fashion: the dark side of fashion exhibition* (21 July–9 September 2017). Workshop participants were immersed in a VR environment using *HTC VIVE* hardware and the creative drawing tool, *Google Tilt Brush*. Participants were instructed to; first, accustom themselves to the VR (*VIVE*) environment, second, to learn basic (*Tilt Brush*) functions and third, to select a 3D database avatar and create a virtual garment. Workshops were open for public participation and no prior VR experience was necessary. Findings suggest the VR domain has potential as a viable Human–Computer Interaction (HCI) domain for both creative and technical apparel design and applicable to a diverse body range.

Keywords Virtual reality · VR · Human–Computer interaction · HCI
Clothing size · Anthropometric · Body scanning

1 Introduction

This paper outlines practice-led preliminary research into the use of virtual reality (VR) as an immersive design environment for conceptual and technical development of apparel, be this for fashion, or functional applications such as uniforms or protective clothing.

K. Kennedy (✉)
RMIT University Melbourne, Melbourne, Australia
e-mail: kate.kennedy@rmit.edu.au

© Springer Nature Switzerland AG 2019
W. K. Wong (ed.), *Artificial Intelligence on Fashion and Textiles*,
Advances in Intelligent Systems and Computing 849,
https://doi.org/10.1007/978-3-319-99695-0_5

The approach for this investigation has incorporated the use of digital artefacts derived from 3d body scan data, to demonstrate practice approaches to evoke both a technical and cultural discussion of the non-standardised body, i.e., a body model that has not been created as a composite from the numerical statistical average [1] of an anthropometric data set or ad hoc derivative. The prime aim of this research is to investigate approaches that depict a diverse range of body shapes, in physical and virtual formats, to disrupt the questionable statistically average, standard or ideal 'fashion' body. Workshops conducted at the *Slow Fashion Studio* during the, *RMIT Gallery's/Goethe-Institut's* exhibition *Fast Fashion: The dark side of fashion (21 July–9 September 2017)* [2], as a local Melbourne response to the *Fast Fashion* exhibition at the RMIT Gallery, provided the perfect workshop venue for public engagement in experimental (VR) approaches to the production of fashion and apparel for 'everybody'.

2   Background

Mass-produced fashion apparel is produced by a system of standardised sizing by replicating the 'average' or 'ideal' figure type. This approach influences both design and manufacturing modes and prefers limited anthropometric variations to function viably. Sample garments are developed in a base or sample size and then additional larger or smaller sizes are expanded or reduced from the sample size according to the merchandise model stock keeping unit (SKU's) market requirements. The concept of an "average" size is well ingrained within the technical development and production of mass-produced garments. Common tools of the trade such as standard measurements, mannequins, and pattern drafting methods have at their heart the notion of the "average" as the benchmark. The commercial imperative prefers to maintain an idealised average based on the aesthetic of "thin" [3] BMI 18–20. Thus, the fashion ideal, with a preference to thin (=normal) has influenced customer perceptions of clothing sizing. The topic of inconsistently sized clothing is widely discussed in the popular media. Consumers and the apparel industry acknowledge that apparel sold both in store and online does not conform to a size standard and is therefore problematic [4]. Consumer demand is strong for consistent and contemporary clothing size standards. User-friendly online size selection tools continue to be a work in progress.

The business of fast fashion, currently the dominant operational global model for mass-produced apparel is characteristic of; hyper consumption of lower priced goods, high stock turnover, low production costs, and no requirement for original design or product innovation. Over the past 20 years the fashion industry has transitioned from the seasonal indent wholesale supply driven business model, to a commodified retail demand driven system. While the transition to a global mode of fast fashion distribution has greatly expanded the consumer base and interest in fashion product [5], the speed and volume in which the product development and production cycle operate, restricts opportunities for design and product innovation through minimal available development lead time. The current method of producing clothing *en masse* described by Edwards [6] as 'templating' is a major cause of fit issues. This cookie cutter model required to achieve viability for mass production will never satisfy the complexities of the fashion system and can never 'fit vast sections of the population' [7].

Thus, the current model of apparel production, based on a system that requires conformity to a limited range of size options, can never satisfy the anthropometric complexities of the population. Opportunities for innovation in this area, side lined by the expediency of the fast fashion model has limited the capacity for apparel designers to create and design beyond the ideal or average figure and has not provided the incentive to question if the ideal figure (thin) is in fact representative of the actual customer.

The *Fast Fashion* Exhibition curated by Dr. Claudia Banz at the Museum für Kunst and Gewerbe (Hamburg 2015) was previously exhibited in the Philippines (Quezon City 2016) and Indonesia (Jakarta 2017). The *Fast Fashion/Slow Fashion* collaboration, the Australian contribution to the *Goethe-Institute* project *IKAT/eCUT* exhibited alternative approaches to how fashion is produced within a global context. As a provocation the *Fast Fashion* Exhibition invited consideration of broader questions such as: What if clothes were designed in a responsive way that allow for personal variations and preferences? Would you value your clothes more if you had it designed specifically for your body? The *Slow Fashion Studio* provided the venue to test public engagement and reaction to the avatar as a body model within VR.

3   Workshops

The aim of the *Everybody* workshop during the *Slow Fashion Studio* was to test user engagement and reaction (HCI) to first; the VR domain and second; the reaction to digital avatars derived from 'real' 3d body data. The workshop as preliminary research explored the potential for VR technologies to provide an immersive digital 3d environment for both fashion designers and consumers to experience the (hidden) relationship between an individual (non-averaged) body and garment design ideation. The *Every-Body* workshop was the next iteration in explorative creative practice-led research in 3d digital design for fashion and textiles that incorporate critical, discursive, and evaluative aspects to provoke consideration of body assumptions of the dominant mass production (fashion) system. Within the VR environment, a place that fashion design (methods) might occupy in the future, de-identified 3D body scan avatar files become virtual mannequins for potential design/pattern/block development and again invite the consideration of anthropometric diversity, in any scale.

The avatar data files (.obj) or body model database for the *Everybody* workshop, sourced data from two previously exhibited creative works. The first iteration was fifty 1/16th scale 3d printed figurines exhibited as the *Body Collective* in *The Future is Here* RMIT Design Hub, August 2014 [8] and second as *Confronting Morphology* installed in Santa Croce, Florence, *Polimoda, IFFTI conference 2015* [9]. The figurines were exhibited in the RMIT gallery during the *Everybody* workshops as physical representation of the virtual avatars.

Using source data (a figure) from *The Body Collective* digital files, workshop participants were immersed in the VR environment using *HTC VIVE* hardware and the creative drawing tool of the *Google Tilt Brush*. Participants were instructed; to first accustom themselves to the VR (*VIVE*) environment, second to learn basic (*Tilt Brush*) functions, and third to select a 3D database (*Body Collective*) avatar and create a virtual garment. Each workshop session was approximately 20 minutes. Workshops were open for public participation and no prior VR experience was necessary. Workshop protocols were approved by the RMIT Human Research Ethics Committee.

4 Findings

As previously explained the primary aim of the workshop was as a preliminary test of human–computer interaction in VR using *Google Tilt Brush* as a creative tool while interacting with a 3d avatar created from 'real' body scan data.

As both the *Body Collective* and *Confronting Morphology* comprised physical artefacts of 16th scale 3d print figurines, the capacity to scale and interact in full scale with a virtual representation of the figures was the unique research component facilitated via the workshop format. Observing participant response and capacity to interact in VR was the primary research objective. According to RMIT Human Research Ethics Committee approved workshop participation protocol, participants were advised on operating procedures before commencing the VR session. Protocol included; information on the VR environment, the instruction to stop if feeling unwell or dizzy (VR immersion can induce motion sickness), the exit survey process, what information is being recorded and how the information may be used.

The first question asked at the end of the VR session was to check if participants were feeling ill or dizzy. Of the twenty-four (24) participant responses to the question "*how are you feeling?*", twenty-two (22) had an overwhelmingly positive response, with one, "*feeling a little sick, but did not need to stop*" and one "*feeling a little dizzy*" (Fig. 1). None of the participants asked to stop, with the majority having to be asked to conclude the session. Eight (8) (33.3%) of participants had tried VR before and sixteen (16) (66.7%) of the participants were first-time users.

The majority of users were from the discipline of textile and fashion design (Fig. 2). When asked to rank "did you enjoy the experience?" Likert scale (1 = no – 5 = yes), 83.3% of (20) participants scored at level five, with three participants euphorically asking to score five plus, and 16.7% of (4) participants scoring at level four.

When asked the question "did the VR environment inform you on body shape diversity in any way?", on a scale (1 = no – 5 = yes), 87.5% of (21) participants scored level 5, and 12.5% of (3) participants scored at level 3.

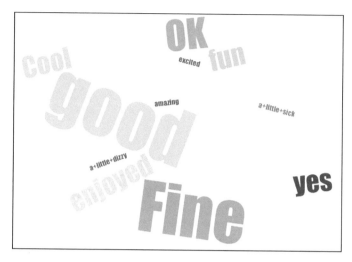

Fig. 1 Question: How are you feeling?

Fig. 2 Question: What is your discipline area?

Anecdotal and concluding remarks revealed very positive and engaged responses. All participants were able to quickly gain mastery over the *Tilt brush* drawing tools. There was an overwhelming positive response to the workshops with a high level of engagement, with participants expressing and discussing applications for textile and apparel development opportunities, such as (Fig. 3.)

"…liked the inside out perspective for example looking at a seam, a pre-cutting conceptual idea space"

"…good way to communicate with your designer/clothing maker"

"…really interesting and helpful, helpful for patterns for a whole body and proto-typing"

"…good to be immersed, a type of mindfulness, engaging otherworldly like"

Fig. 3 Question: Ad hoc responses to VR experience?

"…I have never draped on a body in digital, can see applications into Rhino to experience the form, I think there is potential, gives a different level, mixes sketching by hand and CAD"

"…I chose Edna and I like the randomness of the selection process and not to have to select a body type, I feel I would have selected a fashion stereotype, opportunity to bring to life, added a human element and can see commercial applications, visualising on a person is a big help in the commercial world"

"…drawing in 3D was good, could be interesting to bring sketches into 3D to practice for pattern and print".

5   Conclusion

The response to *Everybody: Immersive Fashion Design*, was a strong indication of positive participant engagement using HCI in VR. Content created by participants and ideas expressed in the exit survey, covered a wide range of applications, from personal fit to design resolution. As an initial test in introducing inexperienced participants to a 3d real time creative space in VR, considering the limited workshop time frame of 20 min, the overwhelmingly positive response and the digital archive of work created, provides a range of opportunities for further research and development for *Everybody: Immersive Fashion Design* in VR.

References

1. Cryle, P., Stephens, E.: Normality: A Critical Genealogy. The University of Chicago Press, Chicago, London (2017)
2. Tsitas, E.: Beyond Fast Fashion: Imagining a Future Without Fashion Waste. https://www.rm it.edu.au/news/all-news/2017/jul/imagining-a-future-without-fashion-waste. Accessed 16 May 2018
3. Sterns, P.: Fat History: Bodies and Beauty in the Modern West, NYU Press (2002)
4. Browne, K.: *CHOICE* [Online] (2014). Available: http://www.choice.com.au/reviews-and-tests/ food-and-health/beauty-and-personal-care/clothing/clothing-size-irregularities.aspx. Accessed 10 May 18
5. Aedy, R.: Economics of fashion. In: Aedy, R (ed.) The Money. ABC Radio National. (2017) http://www.abc.net.au/radionational/programs/themoney/economics-of-fashio n/8494040. Accessed 16 May 2018
6. Edwards, T.: Fashion in Focus: Concepts, Practices and Politics, Milton Park, Abingdon. Oxon, New York, Routledge (2011)
7. Edwards, T.: Fashion in Focus: Concepts, Practices and Politics, Milton Park, Abingdon, p. 212. Routledge, Oxon; New York (2011)
8. Kennedy, K.: The Body Collective. RMIT Design Hub: The Future is Here (2014) http://desig nhub.rmit.edu.au/exhibitions-programs/the-future-is-here-
9. Kennedy, K.: Confronting Morphology. Santa Croce Florence: Momenting the Memento, Polimoda IFFTI Conference. (2015) https://www.youtube.com/watch?v=N-WgeuVhIwQ&lis t=PLpnIQcVnXW5KTeVoOKf89Vk3HCVvippDG&index=2 (5:43secs)

Fabric Defect Detection Based on Faster RCNN

Bing Wei, Kuangrong Hao, Xue-song Tang and Lihong Ren

Abstract Considering that the traditional detection of fabric defect can be time-consuming and less-efficient, a modified faster regional-based convolutional network method (Faster RCNN) based on the VGG structure is proposed. In the paper, we improved the Faster RCNN to suit our fabric defect dataset. In order to reduce the influence of input data for Faster RCNN, we expanded the fabric defect data. Meanwhile, by taking the characteristics of the fabric defect images, we reduce the number of anchors in the Faster RCNN. In the process of training the network, VGG16 can extract the feature map through the 13 conv layers in which the activation function is Relu, and four pooling layers. Then, the region proposal network (RPN) generates the foreground anchors and bounding box regression, and then calculates the proposals. Finally, the ROI Pooling layer uses the proposals from the feature maps to extract the proposal feature into the subsequent full connection and softmax network for classification. The experimental results show the capability of fabric defect detection via the modified Faster RCNN model and indicate its effectiveness.

Keywords Fabric defect · Classification · Detection · VGG16 · Faster RCNN

1   Introduction

In most textile mills, the visual detection of trained workers is still a critical element in the fabric defect inspection process with low detection precision and efficiency subjected to different fibers, visual fatigue and many other constraints. At present, there are several types of fabric defect detection algorithms: (1) statistical-based algorithms, e.g., co-occurrence matrix [1], morphology [2]. (2) spectral-based algorithms, transform the image information from the time domain to frequency domain,

B. Wei · K. Hao (✉) · X. Tang · L. Ren
Engineering Research Center of Digitized Textile and Apparel Technology,
Ministry of Education, College of Information Sciences and Technology,
Donghua University, Shanghai 201620, People's Republic of China
e-mail: krhao@dhu.edu.cn

© Springer Nature Switzerland AG 2019
W. K. Wong (ed.), *Artificial Intelligence on Fashion and Textiles*,
Advances in Intelligent Systems and Computing 849,
https://doi.org/10.1007/978-3-319-99695-0_6

e.g., Gabor Transform [3, 4], Fourier Transform [5]. (3) model-based algorithm, the random filed of image is a stochastic model of random variables. Autoregression model [6] belongs to one-dimensional model. (4) Learning-based algorithms, e.g., neural network [7–9], support vector machines [10]. These methods can further extract the feature information of the defective image. CNN is the deep neural network model [11, 12], which is inspired from the human visual mechanism. With the characteristics of CNN, Girshick et al. designed region-based convolutional neural network (RCNN) [13], obtained the candidate region by region selection approach. Then, they presented a Fast RCNN approach [14] for object detection. What's more, after the accumulation of RCNN and Fast RCNN, their team proposed Faster RCNN [15] for object detection.

In this paper, in order to reduce the influence of model overfitting in the training phase, we augment the fabric defect data. At the same time, because each fabric defect image in our fabric defect dataset has only one defect, we reduce the number of anchors in the Faster RCNN.

The remainder of the paper is organized as follows: Faster RCNN network is overviewed in Sect. 2; fabric defect detection based on Faster RCNN is introduced in Sect. 3; some experiment design and results are contained in Sect. 4, and Sect. 5 concludes the paper.

2 Overview of Faster RCNN

When the RCNN and Fast RCNN were proposed, Girshick et al. presented Faster RCNN [15] that integrated feature extraction, regional proposal generation, rectification in the model as shown in Fig. 1. The structure of Faster RCNN consists of convolution layers, RPN, ROI pooling and classification.

2.1 Convolution Layers

As the image classification and detection method, CNN can extract the feature maps [16]. In the study, we use the VGG16 as the CNN model for feature extraction.

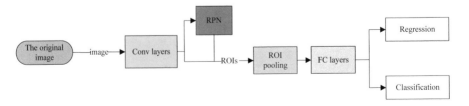

Fig. 1 Faster RCNN structure

Firstly, the input is the fabric defect images, and image feature maps are extracted by convolution layer with Relu activation function and pooling layers [17].

2.2 Region Proposal Networks

As shown in Fig. 2, the center point of the 3×3 sliding window can be mapped back to the center point of the original image. For each anchor, the output of classification layer is the score, which represent the region of interests or the background. The output of regression layer has four coordinates that indicate the coordinate position of the fabric defect.

2.3 ROI Pooling and Classification

The region proposals and input feature maps can be collected through ROI pooling layer. ROI pooling layer is characterized by the non-fixed size of feature maps. The output of ROI pooling layer is the vector, whose size is channel $* w * h$, where w and h are the width and height of the feature map, and channel means the dimension of the feature map. Classification layers calculate the proposal's class by using the full connection layer and softmax network. Bounding box regression is to get the final exact position and return more accurate target detection box.

Fig. 2 Region proposal network (RPN)

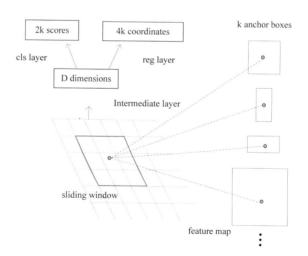

3 Fabric Defect Detection Based on Faster RCNN

3.1 Data Preparation and Augment

In this paper, our fabric defect data is collected from the textile mills. We detect six types of fabric defects that are brokenpick, felter, drawback, sundries, brokenend, and oilstians. In order to reduce the influence of model overfitting, we augment the fabric defect dataset with horizontal and vertical flips. The images with horizontal and vertical flips are shown in Figs. 3.

3.2 Reduce the Anchors

In our fabric defect image data set, we find that some fabric defects are small and slender. Meanwhile, there is only one defect on each fabric defect image. In this paper, the anchors were modified by considering the characteristics of fabric defect image. Namely in the original Faster RCNN model, a total of $9 * H * W * D$ candidate proposal regions are predicted, where H and W are the height and width of the feature map, and D means the dimension of the feature map. We only need $6 * H * W * D$ candidate proposal regions are predicted for the feature graph of the model output in this work.

4 Experiment Design and Results

4.1 Data Creation

The fabric defect dataset is built by ourselves. The dataset, partially presented in Fig. 4, has six different defect classes of images. The flaw categories are brokenpick, felter, drawback, sundries, brokenend, oilstians. The number of fabric defect images in each class is 135. The size of each image is 227×227.

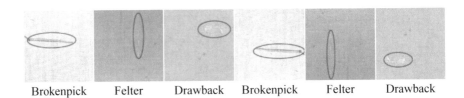

Brokenpick   Felter   Drawback   Brokenpick   Felter   Drawback

Fig. 3 Vertical flips and horizontal flips of the images

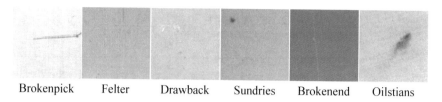

| Brokenpick | Felter | Drawback | Sundries | Brokenend | Oilstians |

Fig. 4 Different types of the fabric defects in the dataset

4.2 Experiment Results

In this section, the following detection results are obtained by modified Faster RCNN. The examples of testing results of the model to fabric defect detection are shown in Fig. 5.

In Fig. 5, we just show some examples of fabric defect detection results. It can be seen that the modified Faster RCNN can detect the fabric defects accurately.

During the training process, the time cost of training the modified Faster RCNN is 617.52 s. Table 1 shows time-consuming of fabric defect detection. We can see that the average detection time is about 0.3 s for each type of fabric defects. We can achieve near real-time detection rates using the model. Performance comparison of different detection approaches is identified by Table 2. Nums represents the number of defect type. Accuracy represents the fabric defect detection accuracy. We can see from the experimental results that the modified Faster RCNN outperforms the other

Fig. 5 The examples of fabric defect detection

Table 1 Time-consuming of the fabric defect detection

Defect images	Sundries	Oilstains	Brokenpick	Felter	Drawback	Brokenend
Testing time (s)	0.26	0.308	0.269	0.259	0.288	0.283
	0.304	0.294	0.282	0.267	0.271	0.264
	0.279	0.344	0.276	0.270	0.278	0.278
	0.299	0.259	0.280	0.283	0.283	0.268
	0.296	0.268	0.271	0.285	0.266	0.267
Average time (s)	0.287	0.294	0.275	0.272	0.276	0.272

Table 2 Performance comparison of different detection approaches

Model	MLBP [18]	Gabor filter [19]	Nonlocal sparse [20]	Ours
Nums	6	4	4	6
Accuracy (%)	83.6	92.3	94.1	**95.8**

Accuracy represents the fabric defect detection accuracy that indicates the percentage of correct classification of all testing samples

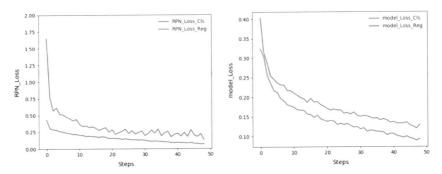

Fig. 6 The results of loss in the RPN network modified Faster RCNN

existing methods. Figure 6 show the results of loss in RPN network and the modified Faster RCNN. It can be seen that the model achieves low loss during the model training. The minimum loss of classification and regression loss in RPN network are 0.119 and 0.088, respectively. The minimum loss of classification and regression loss in modified Faster RCNN are 0.132 and 0.065, respectively.

5 Conclusion

In this paper, the modified Faster RCNN solves the problem of extracting the feature maps by means of CNNs. In consideration of the characteristic of the fabric defect images, the Faster RCNN has been modified to suit our fabric defect data set. Through the modified Faster RCNN, we can get high detection accuracy and achieve near real-time detection rates. However, with the help of deep learning, the detection method is more effectively compared with traditional approaches, but the training of the Faster RCNN can be a little more time-consuming. Meanwhile, the detection model may select more parts of the object in the bounding box regression, we can improve it to be more intuitive in the further.

References

1. Zhu, D., Pan, R., Gao, W.: Yarn-dyed fabric defect detection based on autocorrelation function and GLCM. Autex Res. J. **15**(3), 226–232 (2015)
2. Mak, K.L., Peng, P., Yiu, K.F.C.: Fabric defect detection using morphological filters. Image Vis. Comput. **27**(10), 1585–1592 (2009)
3. Zhu, Q., Wu, M., Li, J.: Fabric defect detection via small scale over-compete basis set. Textile Res. J. **84**(15), 1634–1649 (2014)
4. Jia, L., Chen, C., Liang, J.: Fbaric defect inspection based on lattice segmentation and Gabor filtering. Neurocomputing **238**, 84–102 (2017)
5. Hu, G., Wang, Q., Guo, H.: Unsupervised defect detection in textiles based on Fourier analysis and wavelet shrinkage. Appl. Opt. **54**(10), 2963–2980 (2015)
6. Cohen, F.S., Fan, Z., Attali, S.: Automated inspection of textile fabrics using textural models. IEEE Trans. Pattern Anal. Mach. Intell. **13**(8), 803–808 (1991)
7. Li, Y., Zhao, W., Pan, J.: Deformable patterned fabric defect detection with fisher criterion-based deep learning. IEEE Trans. Autom. Sci. Eng. **14**(2), 1256–1264 (2017)
8. Kumar, A.: Neural network based detection of local textile defects. Pattern Recogn. **36**(7), 1645–1659 (2003)
9. Huang, C.C., Chen, I.C.: Neural-fuzzy classification for fabric defects. Text. Res. J. **71**(3), 220–224 (2001)
10. Li, W., Cheng, L.: yarn-dyed woven defect characterization and classification using combined features and support vector machine. J. Textile Inst. Proc. Abstr. **105**(2), 163–174 (2014)
11. Simonyan, K., Zisserman, A.: Very deep convolutional networks for large-scale image recognition. In arXiv: 1409.1556, Sept 2014
12. Yann, L., Bengio, Y., Hinton, G.: Deep learning. Nature **521**(7553), 436–444 (2015)
13. Girshick, R., Donahue, J., Darrell, T., Malik, J.: Region-based convolutional networks for accurate object detection and segmentation. IEEE Trans. Pattern Anal. Mach. Intell. 38(1), 142–158, January
14. Girshick, R.: Fast R-CNN. IEEE International Conference on Computer Vision, pp. 1440–1448 (2015)
15. Ren, S., He, K., Girchick, R., Sun, J.: Faster R-CNN: towards real-time object detection with region proposal networks. IEEE Trans. Pattern Anal. Mach. Intell. **39**(6), 1137–1149 (2017)
16. Bell, S., Zitnick, C.L., Bala, K., Girshick, R.: Inside-outside net: detecting objects in context with skip pooling and recurrent neural networks. In: IEEE Conference on Computer vision and Pattern Recognition, pp. 2874–2883 (2016)
17. Szegedy, C., Liu, W., Jia, Y., Sermanet, P., Reed, S., Anguelov, D., Erhan, D., Vanoucke, V., Rabinovich, A.: Going deeper with convolutions. In: IEEE Conference on Computer Vision and Pattern Recognition, pp. 1–9 (2015)
18. Liu, Z., Yan, L., Li, C.: Fabric defect detection based on sparse representation of main local binary pattern. Int. J. Cloth. Sci. Technol. **29**(3), 282–293 (2017)
19. Tong, L., Wong, W.K., Kwong, C.K.: Differential evolution-based optimal Gabor filter model for fabric inspection. vol. 173, pp. 1386–1401, January, 2016
20. Tong, L., Wong, W.K., Kwong, C.K.: Fabric defect detection for apparel industry: a nonlocal sparse representation approach. IEEE Access **5**, 5947–5964 (2017)

Woven Light: An Investigation of Woven Photonic Textiles

Lan Ge, Jeanne Tan, Richard Sorger and Ziqian Bai

Abstract Textile materials, which have been widely used in apparel, interior design, health care, automobile, and infrastructure, play a fundamental role and exhibit a close relationship to human's daily life. For the past few decades, smart textile materials with new performance, functionality, and design values have been developed through the integration of advanced technologies [1]. This study aims to review and examine current approaches to realize photonic textiles, including LED-embedded, photoluminescent, and polymeric optical fiber (POF) textiles. This work also reviews examples of woven interactive textiles and discusses how weave parameters can affect POF textiles. POF textiles, which exhibit the advantage of directly incorporating functional fibers during the fabrication process, have the potential to serve as a flexible textile display. In addition, this material provides high design values due to its dynamic visual appearance. One of the fabrication methods used to produce POF textiles is the weaving technique. The properties of smart textiles, including hand feel, elasticity, flexibility, and drapability, can be considerably changed through the interplay among weaving variables, such as fiber content, patterns, weave structures, fabric construction, and finishing techniques. Complex woven structures can be designed to simplify the production process; novel yarns can be used to provide variety to textile appearance and tactility. However, few studies have been conducted on the fabrication process of smart textiles through weaving. Tan and Bai developed prototypes of interactive cushions using POF textiles integrated with touch sensors [2, 3]. The design process of these prototypes focuses on the integration of electronic components and interactivity instead of on the textile fabrication method. A literature review indicates that research on the weaving structures of POF textiles is lacking. The use of weaving techniques to create innovations for smart textiles to enhance their appearance, performance, and functionality demonstrates potential.

L. Ge · J. Tan (✉) · Z. Bai
Institute of Textiles and Clothing, The Hong Kong Polytechnic University,
Hung Hom, Hong Kong
e-mail: jeanne.tan@polyu.edu.hk

R. Sorger
School of Design, Kingston School of Art, London, UK

© Springer Nature Switzerland AG 2019
W. K. Wong (ed.), *Artificial Intelligence on Fashion and Textiles,*
Advances in Intelligent Systems and Computing 849,
https://doi.org/10.1007/978-3-319-99695-0_7

Keywords Interactive photonic textile · Weaving · Textile design

1   Introduction

Textile materials, as the most intimate and fundamental materials in people's daily life, have achieved new performance, functionality, and design values during the past few decades through the integration of advanced technologies [1]. In particular, smart photonic textiles with the capability to display illuminative colors, patterns, and texts from light emission have been developed.

This study reviews and examines different approaches to achieve photonic textiles, including light-emitting diode (LED)-embedded, photoluminescent, and polymeric optical fiber (POF) textiles.

2   Smart Photonic Textiles and Clothing

2.1   *LED-Embedded Textiles*

Switch Embassy, a New York-based wearable technology company, created a smart shirt that displays customized illuminative text messages [4]. The shirt is embedded with LED textile panels. It allows users to instantly upload animations, static text, scrolling text, and graphics onto the fabric from the mobile apps on their smartphones via Bluetooth (Fig. 1). The smart shirt creates a new platform for self-expression and social interaction through technology-infused fashion.

LED-embedded textiles, which have the capability to display changeable illuminative graphics, have drawn considerable attention from the smart textile industry. This technology has been applied to enhance aesthetics, visualize ideas, and function as a communicative display. However, the main drawback of this technology is that LEDs typically require to be connected in chains. "The failure of a single component in the ring will cause all lights in the ring to go out" [5]. Regardless of the progress in the miniaturization of LED components, existing LEDs remain in the millimeter scale, which causes bulkiness in fabric and makes it uncomfortable to wear.

2.2   *Photoluminescent Textiles*

Ying Gao, a Montreal-based fashion designer, created two interactive dresses called "(No)where (Now)here" which were inspired by the aesthetic of disappearance. The dresses are made from lightweight super organza and can slowly move as the audience gazes at them [6]. Interaction is enabled using an eye-tracking technology. The fabric

Fig. 1 LED-embedded shirt provides an illuminative graphics display

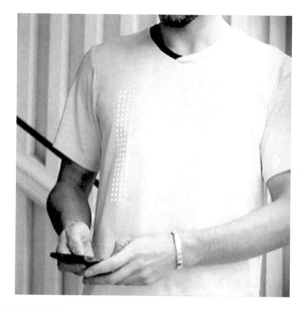

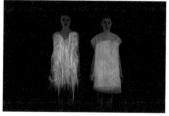

Fig. 2 Interactive dresses designed by Ying Gao

consists of photoluminescent threads that can light up under darkness (Fig. 2). The dress becomes a dynamic interactive platform that connects the artifact, the audience, and the environment.

Photoluminescent textiles are fabricated with yarns that are processed by mixing polyester chips with photoluminescent pigments. These textiles do not require any external electronic components, such as a light source and a power supply. However, they should be exposed to a light source for a certain period to be able to emit light in the dark. In general, sunlight exposure for 1 h will enable the yarn to provide continuous light emission for 3 h. Moreover, photoluminescent textiles only emit light in the dark. They cannot light up on their own, and their light cannot be controlled. Their luminous appearance is also limited because only a few colors of photoluminescent threads are available.

Fig. 3 POF textile cushions

2.3   Polymeric Optical Fiber (POF) Textiles

Prototypes of interactive cushions were developed by Tan and Bai using POF textiles [2, 3]. These textiles can emit light by connecting optical fiber bundles to a LED light source. As the user touches different areas of the cushion, these areas will light up and change colors. Interaction is enabled by touch sensors and a microprocessor unit that are placed inside the cushion. Figure 3 shows the interactive cushion prototypes.

With the advantages of good hand feel, flexibility, lightweight, and drapability, the qualities of POF textiles are similar to those of conventional fabrics, which is crucial for its usability and wearability. POF textiles also require low cost and low power supply. They are produced by directly integrating POF threads into the textile fabrication process, thereby providing unlimited possibilities in terms of surface design and tactility. However, POF can be susceptible to damages from severe bending. Fabrication methods, such as knitting and embroidery, require looping yarns, which can damage POF due to severe bending. By contrast, woven textiles are formed by interlacing yarns with one another. This process involves less bending, and thus causes less damage to POF. Therefore, weaving technology is more suitable for producing POF textiles.

3   Woven Smart Textiles

As one of the traditional textile fabrication methods, weaving exerts considerable impact on the overall performance and functionality of textiles by directly integrating functional fibers into a material. Weaving factors, including fiber content, patterns, woven structure, fabric construction, and finishing techniques, will individually and collectively affect the performance of smart textiles. Many woven structures have not yet been fully explored by researchers. Thus, further innovation in woven structure design for smart textiles demonstrates potential. This section presents recent examples of smart textiles created via the weaving technique.

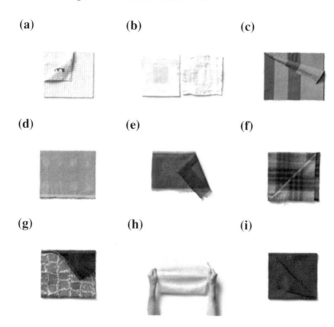

Fig. 4 Interactive Google Jacquard textiles with different properties

3.1 Woven Smart Textiles with Conductive Fibers

Google developed interactive digital textiles that function as unobtrusive and effective interface for computers [7]. This novel textile material enables users to control their phone wirelessly through interactivity with the textile, such as multi-touch and gesturing over the surface. These interactive textiles are manufactured at scale using weaving technology. Novel conductive yarns are woven in warp and weft directions, thereby combining with substrate materials. Different fabric properties, including elasticity, tactility, and multilayers, can be achieved in interactive textiles through variation and interplay among weave structures, patterns, and materials (Fig. 4). The connection between smart textiles and electronics has also been significantly simplified through an advanced weave structure that combines two layers and floats.

3.2 Woven Smart Textiles with Photonic Band Gap (PBG) Fiber

XS Labs [8, 9] developed textiles woven with PBG Bragg fibers on Jacquard and hand looms. Unlike other optical fibers, PBG Bragg fibers display iridescent colors under ambient illumination even without abrading the fibers. The colors appear differently from ambient side illumination and guided light transmission from the end of the

Fig. 5 PBG textiles woven
on a Jacquard loom

fibers. The design group experimented with weave structures to achieve different
visual appearance of photonic textiles (Fig. 5). However, textile colors, patterns, and
textures remain limited and require further explorations. Compared with hand looms,
Jacquard looms provide more design varieties and more dramatic changes to fabric
properties and appearance.

4   Discussion and Conclusion

The unique illuminative quality of interactive photonic textiles provides a dynamic
visual appearance and potential to function as a flexible communicative display. With
the advantages of lightweight, flexibility, and low cost, POF is an ideal material for
developing interactive photonic textiles by being directly integrated into textiles.
During the smart textile fabrication process, different weave parameters can affect
textile properties, functionality, and connection with electronic components [7]. The
production process can be simplified by balancing woven structures. Many varieties
of smart textile appearance and tactility can also be achieved by combining novel
yarns and complex weave structures. However, the development of existing POF
textiles has been focused on electronics integration, product-making process, and
interactive functions, whereas weave experimentations have been neglected. An in-
depth investigation into the relationships between weaving and the performance of
POF textiles should be conducted to achieve innovative solutions. With the interplay
among weaving variables, such as fibers, weave structures, patterns, fabric construction, and finishing techniques, there are infinite possibilities to change the proper-

ties of POF textiles, including their hand feel, elasticity, flexibility, and drapability, thereby enhancing their appearance and functionalities.

References

1. Koncar, V.: Introduction to smart textiles and their applications. In: Smart Textiles and Their Applications, pp. 1–8. Elsevier, New York (2016)
2. Tan, J.: Photonic fabrics for fashion and interiors. In: Handbook of Smart Textiles, pp. 1005–1033. Springer, Berlin (2015)
3. Bai, Z.Q., Tan, J., Johnston, C., Tao, X.M.: Connexion: development of interactive soft furnishings with polymeric optical fibre (POF) textiles. Int. J. Clothing Sci. Technol. **27**, 870–894 (2015)
4. Switch Embassy.: http://switchembassy.com/ (2015)
5. Rowley, E.: Lighting up your garments: an investigation into methods of making fabrics glow. In: CHI 2009, Boston, USA (2009)
6. Gaddis, R.: What is the future of fabric? These smart textiles will blow your mind. http://www.forbes.com/ (2014)
7. Poupyrev, I., Gong, N.W., Fukuhara, S., Karagozler, M., Schwesig, C., Robinson, K.: Project Jacquard: interactive digital textiles at scale. In: the 2016 CHI Conference on Human Factors in Computing Systems, San Jose, California, USA. https://doi.org/10.1145/2858036.2858176 (2016)
8. Berzowska, J., Skorobogatiy, M.: Karma chameleon: jacquard-woven photonic fiber display. In: SIGGRAPH'09, New Orleans, Louisiana. https://doi.org/10.1145/1597990.159800 (2009)
9. Berzowska, J.: http://xslabs.net/karma-chameleon/site/prototypes.php

A Piezoelectric Energy Harvester for Wearable Applications

Wenying Cao, Weidong Yu and Wei Huang

Abstract Mechanical energy can be captured and converted into electric energy directly through piezoelectric energy harvester, which meet the need of portable power supply for wearable electronics. Due to the large amount of power generation from walking or running in daily life, putting the harvester into shoes is a common research method. An efficient energy harvester made of polyvinylidene difluoride (PVDF) film with a zigzag structure is proposed and demonstrated for general human motion in this paper. The designed zigzag structure is a PVDF film layer sandwiched by two stainless steel layers. Two experiments were conducted to study the output voltages generated by the 500 N impulse force under 1 Hz and the human walking varies different speeds. The peak power generated by the impulse force and human walking are all at the μW level. The human test result shows that the peak-peak voltage increases with the increase of the walking speed.

Keywords Wearable · Piezoelectric · Energy harvester

1   Introduction

Over the past ten years, research studies on the conversion of human movement into useful electrical energy based on piezoelectric materials have been activity conducted [1–3]. The piezoelectric energy harvesters can convert mechanical energy into electric energy directly resulted in their structures can be very simple, so many kinds of human movement can be considered as an energy source to be harvested to power wearable electronics. These studies include the energy harvested from the

W. Cao · W. Yu (✉) · W. Huang
Donghua University, Shanghai 201620, China
e-mail: wdyu@dhu.edu.cn

W. Cao
e-mail: 1115133@mail.dhu.edu.cn

W. Huang
e-mail: 120400534@mail.dhu.edu.cn

© Springer Nature Switzerland AG 2019
W. K. Wong (ed.), *Artificial Intelligence on Fashion and Textiles*,
Advances in Intelligent Systems and Computing 849,
https://doi.org/10.1007/978-3-319-99695-0_8

bending of elbow [4], implants in the knee joints [5], motion of the human limbs [6], the throbbing of pulse wave [7], breathing [8], walking [9]. Among these wearable piezoelectric harvesters, the one of harvesting energy from walking perhaps is the most common method due to the large amount of power generated from walking or running in daily life.

The usual way of converting the mechanical energy of walking into electrical energy mainly is to integrate the piezoelectric materials into shoes because of its limited installation area. There are two familiar places to implant the harvester, one is the sole of shoe where can be convenient for the flat plate type and arch type energy harvesters to tap the energy from striking and harness the energy dissipated in bending of the sole [10, 11], another is the cavity of the heel of shoe where can be convenient for the cantilever type to collect the vibrating energy [1].

Lead zirconate titanate (PZT) and polyvinylidene difluoride (PVDF) are the most common piezoelectric materials for energy harvesting. PZT is widely used in power harvesting applications, but it has brittle nature. Although PVDF offers a lower energy conversion efficiency than PZT, but PVDF is more flexible and easier to shape. They have their own advantages and disadvantages, hence, we can choose which one to use according to their properties and the design of the harvester. The harvester proposed in this paper is based on the PVDF due to its flexibility.

This paper developed a zigzag structural piezoelectric energy harvester which is based on a sandwich structure, aiming at harvesting the energy at a low frequency. The harvester is a thin geometrical form, so it can be not only integrated in a shoe readily but also can be embedded to the carpet easily. The performance of the zigzag harvester was tested and evaluated to proving that this structure makes it possible to generate electricity with the low-frequency excitation.

2   Description of the Zigzag Harvester

A simply sketch of the zigzag harvester studied in this paper is shown in Fig. 1a. The main structure of the harvester is a sandwich structure, where a layer PVDF thin film metalized on both sides is sandwiched by two stainless steel layers. An ultrathin insulation layer is put between the PVDF layer and the stainless steel layer. The zigzag structure consists of two identical equicrural triangles. The length of the two equal sides of the equicrural triangle l is 5 mm, and the height h is 2 mm, the l and h are presented in Fig. 1b.

The stainless steel layer is about 200 μm thick, 20 mm long, 20 mm wide. The dimensions of length, width, and thickness of the PVDF are 30, 20, and 0.03 mm, respectively. The other geometric and material parameters of PVDF are listed in Table 1.

The special design of the harvester reduces the harvester thickness, which makes it possible to integrate the harvester into a shoe whose inner space is limited. The harvester can be mounted to the insole of the heel. In addition, when stepping on the harvester, the smaller of the height change, the more comfortable of the foot.

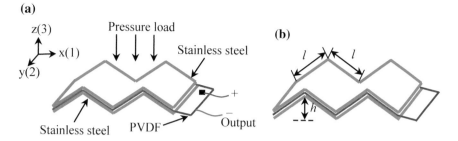

Fig. 1 Sketch of PVDF energy harvester. **a** The sandwich structure of the harvester; **b** The design parameters

Table 1 Material properties of the PVDF

Parameters	Value
d_{33} (PC/N)	21
d_{31} (PC/N)	17
Coupling coefficient k_{33} (%)	10–14
Density (kg/m^3)	1.78×103
Relative permittivity	9.5 ± 1.0
Elastic modulus(MPa)	2500

3 Experiments

3.1 Performance Test

The purpose of this test is to investigate the output voltage generated by the impulse force. The test of the power harvester was based on a simple force excitation generator which can provide an impulse force, as shown in Fig. 2. During testing, the harvester was fixed to the buffer board and the impulse force was set to 500 N. The peak-peak output voltage by the impulse force at a frequency of 1 Hz was measured by the oscilloscope. The internal resistance of the oscilloscope was 10 MΩ, serving as a resistive load.

3.2 Human Test

The harvester was installed in the heel of the shoe due to more force provided by body weight at that area, as depicted in Fig. 3. The power harvester was tested based on the human walking on the treadmill. When the human was walking on the treadmill by wearing the piezoelectric shoe, the output voltage was measured by the oscilloscope. The peak-peak output voltages vary with different walking speed were recorded and

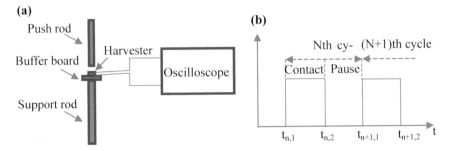

(a)

Push rod

Buffer board

Harvester

Support rod

Oscilloscope

(b)

Nth cy- (N+1)th cycle

Contact Pause

$t_{n,1}$ $t_{n,2}$ $t_{n+1,1}$ $t_{n+1,2}$ t

Fig. 2 Scheme of the simple force excitation generator. **a** Structure; **b** Schematic

Fig. 3 The harvester placed in a shoe

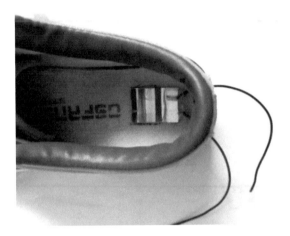

analyzed. The experimental walking speeds were 3, 4, 5, 6 km/h, respectively. When the speed reach or exceed 5 km/h in the treadmill, the ways of brisk walking and slow jogging all can be realized. In this paper, the way of brisk walking was chosen.

4   Results and Discussion

4.1   Performance Test

Figure 4 shows the output voltage waveform generated by the piezoelectric device at a 1 Hz impulse force. It can be seen that the waveform has small fluctuation and it is stable. The peak-peak voltage can reach 5.36 V. It can tell that the corresponding peak power which calculated by the equation $p = U^2/R$ of the harvester is at the μW level.

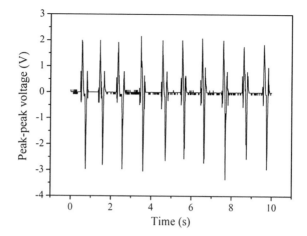

Fig. 4 Output voltage waveform of the zigzag harvester

4.2 Human Test

Figure 5 represents the peak-peak output voltage measured with the same load resistance under different walking speeds. It can be seen that the peak-peak voltage increases with the increased walking speed. The peak-peak voltage can reach 5.032 V at 6 km/h. The frequency of the faster walking is obviously bigger than that of the slower walking, which resulted in the peak-peak voltage increases as the walking speed increases. The faster walking speed lead to greater impulse force in the human heel [12], which is the another reason that the peak-peak voltage of faster walking being larger than that of the slower walking. Hence, if the power is in urgent need, it is very convenient to improve the output performance fastly by increasing the walking speed anytime and anywhere.

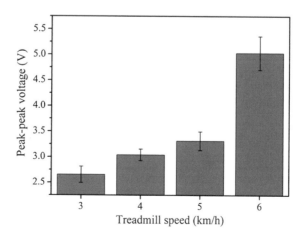

Fig. 5 Comparison of the output voltage versus time for different walking speed

5 Conclusions

The shoe-inserted harvester will bring a lot of conveniences to us because the harvester can harvest energy anytime and anywhere. The main objective of this research is to design a zigzag structural piezoelectric energy harvesting device, consisting of one PVDF film and two stainless steels. About 5.36 V peak-peak voltage was generated under a 500 N impulse force at 1 Hz. While about 5.032 V peak-peak voltage was generated when the human walking on the treadmill at 6 km/h. The results of human test reveal that the peak-peak voltage increases with the increase of the walking speed. Though the output power is not very big, more numbers of the PVDF layer can be stacked to improve the output power in the future work.

References

1. Hwang, S.J., Jung, H.J., Kim, J.H., Ahn, J.H., Song, H., Song, Y., Lee, H.L., Moon, S.P., Park, H., Sung. T.H.: Designing and manufacturing a piezoelectric tile for harvesting energy from footsteps. J. Curr. Appl. Phys. **15**(6), 669–674 (2015). https://doi.org/10.1016/j.cap.201 5.02.009
2. Cao, W.Y., Gu, Q.J., Liu, Y.T., Yu, W.D.: Research status of human energy harvesting. J. Micronanoelectronic Technol. **53**(02), 78–86 (2016). https://doi.org/10.13250/j.cnki.wndz.20 16.02.002
3. Cao, W.Y., Nie, F.H., Huang, W., Gu, Q.J., Yu, W.D.: Current situation and development trend of piezoelectricwearable energy harvesters. J. Electron. Compon. Mater. **35**(8), 6–10 (2016). https://doi.org/10.14106/j.cnki.1001-2028.2016.08.002
4. Yang, B., Yun, K-S.: Efficient energy harvesting from human motion using wearable piezoelectric shell structures. In: 16th International Conference on Solid-State Sensors, Actuators and Microsystems, pp. 2646–2649. IEEE Press, Beijing (2011). https://doi.org/10.1109/trans ducers.2011.5969874
5. Pozzi, M., Aung, M.S.H., Zhu, M., Jones, R.K., Goulermas, J.Y.: The pizzicato knee-joint energy harvester: characterization with biomechanical data and the effect of backpack load. J. Smart Mater. Struct. **21**(7), 075023 (2012). https://doi.org/10.1088/0964-1726/21/7/075023
6. Renaud, M., Fiorini, P., Schaijk, R.V., Hoof, C.V.: Harvesting energy from the motion of human limbs: the design and analysis of an impact-based piezoelectric generator. J. Smart Mater. Struct. **18**(3), 035001 (2009). https://doi.org/10.1088/0964-1726/21/4/049501
7. Yoon, S., Cho, Y-H.: A skin-attachable flexible piezoelectric pulse wave energy harvester. In: the 14th International Conference on Micro and Nanotechnology for Power Generation and Energy Conversion Applications, pp. 012026. IOP Press, Kobe (2014). https://doi.org/10.108 8/1742-6596/557/1/012026
8. Abdi, H., Mohajer, N., Nahavandi, S.: Human passive motions and a user-friendly energy harvesting system. J. Intell. Mater. Syst. Struct. **25**(8), 923–936 (2014). https://doi.org/10.117 7/1045389X13502854
9. Shenck, N.S., Paradiso, J.A.: Energy scavenging with shoe-mounted piezoelectrics. J. IEEE Micro **21**(3), 30–42 (2001). https://doi.org/10.1109/40.928763
10. Kymissis, J., Kendall, C., Paradiso, J., Gershenfeld, N.: Parasitic power harvesting in shoes. In: the Second International Symposium on Wearable Computers, pp. 132–139. IEEE Press, Pittsburgh (1998). https://doi.org/10.1109/iswc.1998.729539

11. Zhao, J., You, Z.: A shoe-embedded piezoelectric energy harvester for wearable sensors. J. Sens. **14**(7), 12497–12510 (2014). https://doi.org/10.3390/s140712497
12. Burnfield, J.M., Few, C.D., Mohamed, O.S., Perry, J.: The influence of walking speed and footwear on plantar pressures in older adults. J. Clin. Biomech. **19**(1), 78–84 (2004). https://doi.org/10.1016/j.clinbiomech.2003.09.007

Surrogate-Based Modeling and Optimization of the Bleach Washing for Denim Fabrics

Wenbo Ke, Jie Xu, Ming Yang and Changhai Yi

Abstract This research is to develop a framework which can be used for surrogate-based modeling and optimization of the bleach washing for denim fabrics. The aim of the proposed framework is to predict performances and minimize costs of the bleach washing. In the framework, a series of surrogate models are first constructed for illustrating the relationships between bleach washing parameters and washing performances by the orthogonal experimental design, and the constructed surrogate models are proved to own an acceptable accuracy. Then, based on the surrogate models, an optimization model is built, whose objective is to minimize costs of the bleach washing under constraints of performance requirements. The optimization model can be treated as an integer nonlinear programming(INLP)problem and solved by the Monte Carlo simulation.

Keywords Surrogate model · Optimization · Bleach washing · Denim

W. Ke · J. Xu
School of Textile Science and Engineering,
Wuhan Textile University, Wuhan, Hubei, China

C. Yi
Technical Research Institute,
Wuhan Textile University, Wuhan, Hubei, China

W. Ke · J. Xu (✉) · C. Yi (✉)
Hubei Center for Jeans Engineering and Technology,
Wuhan, Hubei, China
e-mail: toxujie@163.com

C. Yi
e-mail: ych@wtu.edu.cn

M. Yang
Hubei Xie Feng Textile Co., Ltd., Jingzhou, Hubei, China

© Springer Nature Switzerland AG 2019
W. K. Wong (ed.), *Artificial Intelligence on Fashion and Textiles*,
Advances in Intelligent Systems and Computing 849,
https://doi.org/10.1007/978-3-319-99695-0_9

1   Introduction

Denim garments have received a wide acceptance in the world. In the evolution of denim garments, the manufacturers found that consumers were willing to pay more on prewashed denim garments which are more soft and washed or vintage looks were becoming key elements to lead fashion trends of denim garments, so various washing techniques are developed to cater to consumer markets and the washing gradually become the core of denim finishing. Nowadays, almost countless variations of wash processing are created, some normally used methods among them are summarized by Paul [1] and Kan [2]. However, in most washing techniques, different levels of washing performance measures and cost are subject to engineers' experience instead of scientific methodology, which results in that the quality of washed productions is instability and the production cost is non-optimal.

Some related researches have conducted analysis of washing techniques. Among these researches, most of them revealed the general relationships between inputting parameters and performance measures of some given washing methods [3–5], but the revealed relationships from these researches can be only used for finding the variation tendency of performance measures through changing some inputting parameters, and cannot be applied for accurate prediction. Some other researches were concern about the predictions [6, 7], but most of them have not used constructed model to optimize washing process. Even if some researches take the optimization into considerations, the unconstrained optimization used in them cannot fully conform to actual production.

Therefore, we propose a framework which can be used for performance predictions and optimization of washing techniques, and the bleach washing is chosen as an example in this paper. In the framework, we first construct surrogate models between parameters and performance measures of the given washing technique. Based on the surrogate models, we construct a mathematical model for optimization, the objectives of the optimization can be variable, which depends on the requirements of manufacturers, and multi-constraints are taken into the considerations that is more close to actual practices. The remainder of this paper is organized as follows. Section 2 describes a typical bleach washing process followed by a major task: response surface between bleaching performance measures and inputting parameters. In Sect. 3, we build an optimization model whose objective is to minimize cost of bleach washing under constrains of technical specifications. In Sect. 4, a case study is conducted to prove feasibility of the proposed model and we conclude the paper in Sect. 5.

2   Surrogate Models of Bleach Washing

The surrogate model in this paper is considered as a simplified abstraction of the actual model of bleach washing which is hard to obtain and illustrate. The surrogate model can be a mathematical relation or algorithm representing between the inputs

Table 1 Frequently used ranges of values for the inputs in the actual production

Inputs (units)	Concentration of the hydrogen peroxide (owf) (%)	Temperature (°C)	Treatment time (min)
Range	3–12	50–80	15–30

and outputs. Therefore, in order to construct surrogate models, the inputs and outputs of the bleach washing are firstly confirmed, the details is described in the Sect. 2.1. After that, response surface that is one type of surrogate models is constructed based on the experiment design, which is elaborated and analyzed in the Sect. 2.2

2.1 Bleach Washing

In bleach washing, desizing and bleaching are two basic steps. In order to focus on the influences of bleaching, the desizing step is considered as one aspect of experiment preparations in this paper, thus the same formulation of desizing is applied to all samples. The bleaching step is usually carried out with strong oxidative bleaching agents, and hydrogen peroxide which is selected as the example is a normally used one of them. In practices, the bleach washing effect usually depends on strength of the bleach liquor, temperature and treatment time. Among them, for a given bleach agent, the strength of the bleach liquor can be converted as the concentration of bleach agent in the liquor. Therefore, concentration of the hydrogen peroxide, x_c, temperature, x_t, and treatment time, x_{tt}, are selected as inputs of the surrogate model. The frequently used ranges of values for the inputs are listed in Table 1.

Some performance measures are tested after the bleaching step. K/S value is used to measure the reflectance o f the dyed fiber and usually be considered as an alternative assessment of dye uptake, thus it is an important indicator of judging effect of decoloration. Tensile strength is a frequently used indicator of measuring wearability, it usually be tested as two parts including weft tensile strength and warp tensile strength. On the other hand, softness is also a critical fabric property and related to washing, the bending rigidity of fabrics usually be used to reflect the softness. Therefore, the K/S value, y_{ks}, weft tensile strength, y_{we}, warp tensile strength, y_{wa}, and bending rigidity, y_{br}, are chosen as outputs.

After confirming inputs and outputs, the process of the bleach washing can be conducted as follows:

Step 1 Desize the denim garments which are replaced by denim fabrics in this paper;
Step 2 Rinse the desized fabrics;
Step 3 Conduct hydrogen peroxide bleaching with given concentration, temperature and treatment time;
Step 4 Rinse the bleached fabrics;

Table 2 Orthogonal experimental inputs and outputs

Samples	Inputs			Outputs			
	Concentration x_c (owf)	Temperature x_t (°C)	Treatment time x_{tt} (min)	K/S y_{ks}	Weft tensile strength y_{we} (N)	Warp tensile strength y_{wa} (N)	Bending rigidity y_{br} (cN cm)
S1	3	50	15	11.484	803.9	1121.5	0.4634
S2	6	50	20	10.726	797.56	1109.67	0.4539
S3	9	50	25	9.726	766.98	1082.05	0.4269
S4	12	50	30	8.246	729.47	1043.15	0.3703
S5	6	60	15	10.687	792.58	1109.13	0.4524
S6	3	60	20	11.303	798.35	1116.12	0.4552
S7	12	60	25	7.961	736.97	1050.13	0.3672
S8	9	60	30	9.437	747.15	1060.07	0.3881
S9	9	70	15	9.363	769.59	1082.13	0.4226
S10	12	70	20	7.546	744.39	1047.53	0.3651
S11	3	70	25	10.988	788.69	1104.57	0.438
S12	6	70	30	10.246	757.98	1073.14	0.3989
S13	12	80	15	7.123	701.39	1011.72	0.3588
S14	9	80	20	8.757	742.15	1051.97	0.3846
S15	6	80	25	9.856	760.12	1064.15	0.3972
S16	3	80	30	10.724	759.51	1076.98	0.3969

Step 5   Measure the K/S value, weft tensile strength, warp tensile strength and stiffness of the bleached fabric.

2.2 Response Surfaces Between the Bleaching Factors and Performance Measures

Response surface methodology (RSM), which is one kind of surrogate models, explores the relationships between explanatory variables and response variables. The main idea of RSM is using polynomial model to construct an approximate structure instead of the actual model according to a sequence of designed experiments. Before conducting RSM, the values of the bleaching inputs are confirmed by orthogonal experimental design. In line with actual production, undue precision is inappropriate, thus the selected 16 combinations of inputs are rounded and listed in the columns of 2–4 in Table 2. Then, the chosen parameters are used to conduct hydrogen peroxide bleaching on denim fabrics, which are 100% cotton, following the steps illustrated in Sect. 2.1. All performance measures are listed in the columns of 5–8 in Table 2.

The quadratic response surface model is selected to illustrate the relationship between inputs and outputs, its general form is described in Eq. (1)

Table 3 The samples used for testing

Samples	Inputs			Outputs			
	Concentration x_c (owf)	Temperature x_t (°C)	Treatment time x_{tt} (min)	K/S y_{ks}	Weft tensile strength y_{we} (N)	Warp tensile strength y_{wa} (N)	Bending rigidity y_{br} (cN cm)
T1	9	60	20	9.624	774.15	1082.78	0.4250
T2	12	50	25	8.334	756.15	1061.92	0.3927
T3	6	70	20	10.325	783.45	1092.37	0.4365
T4	9	50	30	9.647	758.92	1072.13	0.4118
T5	6	80	30	9.773	738.14	1051.24	0.3675

$$f(x) = b_0 + \sum_{i=1}^{p} b_i x_i + \sum_{j<1}^{p} b_{ij} x_i x_j \sum_{i=1}^{p} b_{ii} x_i^2 \tag{1}$$

where x_i is the ith independent variable, p represents the number of variables, b_0 indicates the regression intercept and b is the coefficient. In this paper, three inputs are considered as independent variables, and the regression analysis is conducted based on the data listed in Table 2 for each performance measure(output). Thus, there are four equations, which are Eqs. (2)–(5), to describe the relationship between inputs and the K/S value, weft tensile strength, warp tensile strength and bending rigidity, respectively.

$$\begin{aligned} f_{ks}(x) = &-0.0238 \cdot x_c^2 - 0.0006 \cdot x_t^2 + 0.0011 \cdot x_{tt}^2 - 0.0030 \cdot x_c \cdot x_t - 0.0001 \cdot x_t \cdot x_{tt} \\ &+ 0.0034 \cdot x_c \cdot x_{tt} + 0.1662 \cdot x_c + 0.0660 \cdot x_t - 0.0625 \cdot x_{tt} + 10.4502 \end{aligned} \tag{2}$$

$$\begin{aligned} f_{we}(x) = &-0.4961 \cdot x_c^2 - 0.0466 \cdot x_t^2 - 0.1841 \cdot x_{tt}^2 - 0.1483 \cdot x_c \cdot x_t + 0.0631 \cdot x_t \cdot x_{tt} \\ &+ 0.0079 \cdot x_c \cdot x_{tt} + 11.2087 \cdot x_c + 4.7200 \cdot x_t + 2.0748 \cdot x_{tt} + 638.7715 \end{aligned} \tag{3}$$

$$\begin{aligned} f_{wa}(x) = &-0.4209 \cdot x_c^2 - 0.0510 \cdot x_t^2 - 0.1209 \cdot x_{tt}^2 - 0.1695 \cdot x_c \cdot x_t + 0.0406 \cdot x_t \cdot x_{tt} \\ &- 0.0483 \cdot x_c \cdot x_{tt} + 11.6988 \cdot x_c + 5.7177 \cdot x_t + 1.0587 \cdot x_{tt} + 938.7627 \end{aligned} \tag{4}$$

$$\begin{aligned} f_{br}(x) = &-0.00076 \cdot x_c^2 - 0.00002 \cdot x_t^2 - 0.00009 \cdot x_{tt}^2 - 0.00014 \cdot x_c \cdot x_t - 0.00008 \cdot x_t \cdot x_{tt} \\ &+ 0.00021 \cdot x_c \cdot x_{tt} + 0.0164 \cdot x_c + 0.0041 \cdot x_t + 0.0077 \cdot x_{tt} + 0.2668 \end{aligned} \tag{5}$$

In order to measure the accuracy of the obtained models, there are other five groups of data (listed in Table 3) are selected as test samples. Based on the inputs of the test samples and the obtained models, we calculate predicted performances (denoted as f_{ks}, f_{we}, f_{wa} and f_{br}) and made comparisons with the observed values (denoted as y_{ks}, y_{we}, y_{wa}, and y_{br}) of the test samples in Table 4, and the root mean squared error (RMSE) for the predictions, which are listed in Table 4, are also calculated to measure the accuracy of the models.

Table 4 The comparisons of the observed values and predicted values for each model

Samples	$f_{ks}(x)$		$f_{we}(x)$		$f_{wa}(x)$		$f_{br}(x)$	
	y_{ks}	f_{ks}	y_{we} (N)	f_{we}(N)	y_{wa} (N)	f_{wa} (N)	y_{br} (cN cm)	f_{br} (cN cm)
T1	9.624	9.523	774.15	779.7	1082.78	1090.7	0.4250	0.4234
T2	8.334	8.271	756.15	750.3	1061.92	1062.3	0.3927	0.3890
T3	10.325	10.353	783.45	784.9	1092.37	1096.7	0.4365	0.4343
T4	9.647	9.682	758.92	745.4	1072.13	1062.8	0.4118	0.4096
T5	9.773	9.869	738.14	745.5	1051.24	1055.1	0.3675	0.3729
RMSE	0.0714		7.7723		6.0407		0.0033	

3 Optimization Model

The proposed optimization model is described in the Eq. (6). The object function in the model is the total cost, but it subject to the constrains which are technical specifications of bleached fabrics should be achieved.

$$
\begin{aligned}
\text{minimize} \quad & c_c \cdot \frac{x_c - l_{cf}}{l_{cu} - l_{cf}} + c_t \cdot \frac{x_t - l_{tf}}{l_{tu} - l_{tf}} + c_{tt} \cdot \frac{x_{tt} - l_{ttf}}{l_{ttu} - l_{ttf}} \\
\text{subject to} \quad & f_{ks}(x_c, x_t, x_{tt}) = t_{ks} \\
& f_{we}(x_c, x_t, x_{tt}) \geq t_{we} \\
& f_{wa}(x_c, x_t, x_{tt}) \geq t_{wa} \\
& f_{br}(x_c, x_t, x_{tt}) \leq t_{br} \text{ or } f_{br}(x_c, x_t, x_{tt}) \geq t_{br} \\
& l_{cf} \leq x_c \leq l_{cu}, l_{tf} \leq x_t \leq l_{tu}, l_{ttf} \leq x_{tt} \leq l_{ttu}
\end{aligned}
\tag{6}
$$

where the l_{cf}, l_{tf}, and l_{ttf} are the lower limit of x_c, x_t and x_{tt}; l_{cu}, l_{tu}, and l_{ttu} are the upper limit of x_c, x_t, and x_{tt}; c_c, c_t, and c_{tt} are the coefficients, which are related to the cost raised by change of concentration of hydrogen peroxide, temperature, and treatment time, respectively; t_{ks} indicates the K/S value should be achieved; t_{we} and t_{ar} are the minimum weft and warp tensile strength; t_{br} can be minimum or maximum bending rigidity depending on the demands.

The constrains in Eq. 6 are nonlinear, and x_c, x_t and x_{tt} are integers in actual production. Thus, the proposed optimization model can be treated as a integer non-linear programming (INLP) problem. For the INLP, a Monte Carlo approach [8] is introduced to solve it, which is a kind of computation intensive methods based on randomization. It converts the original problem (Eq. 6) to a deterministic problem by assigning random values to uncertain parameters x_c, x_t, and x_{tt} in one loop. After conducting a great quantity of loops, the feasible solution can be obtained.

4 Case Study

In this case, the ranges of x_c, x_t and x_{tt} can be set the same with values in Table 1. The c_c, c_t and c_{tt} in the Eq. 6 are defined as 0.05, 0.2, and 0.75. The K/S value of the bleached denim fabric should achieve 10, the minimum weft and warp tensile strength of the fabric should be 700 and 1050 N, respectively, and the bending rigidity of the fabric should be less than 0.37 cN cm. Thus, the Eq. (6) can be written as below

$$
\begin{aligned}
\text{minimize} \quad & 0.05 \cdot \tfrac{x_c - 3}{9} + 0.2 \cdot \tfrac{x_t - 50}{30} + 0.75 \cdot \tfrac{x_{tt} - 15}{15} \\
\text{subject to} \quad & f_{ks}(x_c, x_t, x_{tt}) = 10 \\
& f_{we}(x_c, x_t, x_{tt}) \geq 700 \\
& f_{wa}(x_c, x_t, x_{tt}) \geq 1050 \\
& f_{br}(x_c, x_t, x_{tt}) \leq 0.37 \\
& 3 \leq x_c \leq 12, \, 50 \leq x_t \leq 80, \, 15 \leq x_{tt} \leq 30
\end{aligned}
\tag{7}
$$

The Monte Carlo method is used to solve Eq. 7 by conducting 100,000 loops. The calculated minimum value of object function is 0.71, and the corresponding values of x_c, x_t and x_{tt} are 6, 79, and 25. The calculation time is 0.228 s by using MATLAB under computer configurations: CPU: i7-6560U@ 2.2 GHz, RAM:8G.

5 Conclusion

In this paper, we propose a framework which can be used for performance predictions and optimization of the bleach washing, and hydrogen peroxide bleaching is chosen as an example. In the framework, we firstly construct the a series of quadratic response surface models between inputs and performance measures of the bleaching process by conducting orthogonal experimental design and regression analysis. The constructed quadratic response surface models can be used for performance predictions and own an acceptable accuracy. Then, based on the response surface models, we build an optimization model whose objective is to minimize cost of bleach washing under constrains of technical specifications. The optimization model can be treated as a integer nonlinear programming (INLP) and solved by a Monte Carlo simulation. A case study has been conducted to prove feasibility of the proposed model. Further research issues may include two aspects, one of them focuses on how to construct more variables and multiple objectives for illustrating bleach washing in the optimization model. The other aspect pays more attention on the investigation on constructing the similar framework in other washing techniques.

References

1. Paul, R.: Denim and jeans: an overview. In: Paul, R. (ed.) Denim, pp. 1–11. Woodhead Publishing, London (2015)
2. Kan, C.W.: Washing techniques for denim jeans. In: Paul, R. (ed.) Denim, pp. 313–356. Woodhead Publishing, London (2015)
3. Du, W., Gosh, R.C., Zuo, D.Y., Zou, H.T., Tian, L., Yi, C.H.: Discoloration of cotton/kapok indigo denim fabric by using a carbon dioxide laser. Fibres Text. East. Eur. **24**(4), 63–67 (2016)
4. Kan, C.W., Yuen, C.W.M.: Effect of atmospheric pressure plasma treatment on the desizing and subsequent colour fading process of cotton denim fabric. Color. Technol. **128**(5), 356–363 (2012)
5. Cheung, H.F., Lee, Y.S., Kan, C.W., Yuen, C.W.M., Yip, J.: Effect of plasma-induced ozone treatment on the colour yield of textile fabric. Appl. Mech. Mater. **378**, 131–134 (2013)
6. Hung, O.N., Song, L.J., Chan, C.K., Kan, C.W., Yuen, C.W.M.: Using artificial neural network to predict colour properties of laser-treated 100% cotton fabric. Fibers Polym. **12**(8), 1069–1076 (2011)
7. Hung, O.N., Chan, C.K., Kan, C.W., Yuen, C.W.M., Song, L.J.: Artificial neural network approach for predicting colour properties of laser-treated denim fabrics. Fibers Polym. **15**(6), 1330–1336 (2014)
8. Ichimura, M., Wakimoto, K.: Adaptive Monte Carlo method for solving constrained minimization problem in integer non-linear programming. In: IFIP Technical Conference on Optimization Techniques, vol. 27, pp. 334–342. Springer, Berlin (1974)

Costume Expert Recommendation System Based on Physical Features

Aihua Dong, Qin Li, Qingqing Mao and Yuxuan Tang

Abstract In this paper, we design a Costume Expert Recommendation System (CERS) based on customers' physical features. First, we obtain images of customers, and use a multi-classifier model based on Support Vector Machine (SVM) to extract physical features of customers. The physical features include four features: skin-color, face-shape, shoulder-shape and body-shape. Second, CERS stores the specific physical feature of customers into the Fact Base of the Expert System. It then stores expert knowledge on costume matching into the rule base in the manner of production rules. Finally, the CERS adopts inference engine, namely, blackboard model algorithms to obtain the recommended costume that suits the physical features of the customer. Therefore, the proposed system provides customers an intelligent costume recommendation strategy in accordance with SVM and Expert System.

Keywords Support vector machine · Expert system · Physical features
Costume recommendation

A. Dong (✉) · Q. Li · Q. Mao
College of Information Science and Technology,
Donghua University, Shanghai 201620, China
e-mail: dongaihua@dhu.edu.cn

Q. Li
e-mail: liqin@mail.dhu.edu.cn

Q. Mao
e-mail: 1042764739@qq.com

Y. Tang
Tandon School of Engineering, New York University,
Brooklyn, NY, USA
e-mail: yt1286@nyu.edu

© Springer Nature Switzerland AG 2019
W. K. Wong (ed.), *Artificial Intelligence on Fashion and Textiles*,
Advances in Intelligent Systems and Computing 849,
https://doi.org/10.1007/978-3-319-99695-0_10

1 Introduction

With the advancement of computer technology, online sales of costumes are growing popular [1]. However, customers usually manually input or select text information to describe their physical features, which will result in incorrect information. Therefore, it is difficult for websites to recommend a suitable costume. Furthermore, customers often buy costumes according to their favorite colors or styles, ignoring their physical features, resulting in poor collocation.

At present, the major achievements of costume recommendation are based on establishing size and 3D simulation [2]. In foreign countries, a recommendation system named the right size adopted the technology of "rosetta stone" to recommend size. The disadvantage is that it only offers online fittings, not specific costume recommendations. In China, Zheng Aihua's team [3] applied the Back Propagation (BP) neural network in size recommendation, but the system did not consider the costume style recommendation. In addition, their recommendation methods are similar to the popular websites such as Amazon, Tmall and Taobao. They simply recommend costumes in accordance with the customer's interests and purchasing behaviors, not considering the physical features and the experts' experience.

In this paper, we propose a Costume Expert Recommendation System (CERS) based on physical features, which uses SVM and expert system to provide customers with costumes matching. The process of the CERS is as follows: first, it acquires the customers' images via man–machine interface; secondly, it preprocesses images, extracts physical features by a Histogram of Oriented Gradient(HOG) [4], and classifies physical features with SVM; it then sets up knowledge base collected from the experts, and represents the experts' matching knowledge with production rules; finally, it provides customers an intelligent and efficient costume recommendation method in accordance with blackboard model reasoning. The CERS uses the SVM multiple classifier as the core technology to automatically obtain the customer's physical features involved skin-color, face-shape, shoulder-shape, and body-shape. It solves the problem of subjectivity and uncertainty caused by the customer's manual input or selected text information, not considering the customer's interests and purchasing behaviors.

2 System Overview

The overall structure of the CERS is shown in Fig. 1. The system is mainly composed of two parts: the right part in the figure is the information acquisition module based a SVM multi-classifier which aims to automatically collect the customer's physical feature; the left part is the costumes recommendation module based an Expert System.

Initially, the customer uploads his (or her) photo in the information acquisition module. This module includes image processing, training sample, feature extraction, and SVM-based classification. After these four steps, the information of skin-color,

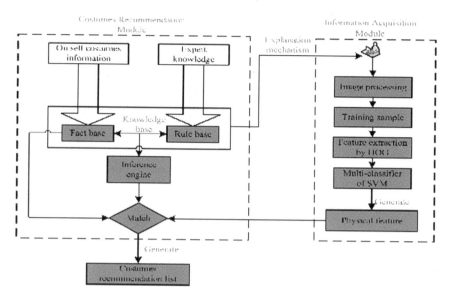

Fig. 1 Structure of the CERS

face-shape, body-shape, and shoulder-shape of the customer is obtained. Additionally, Expert System is engaged to fulfill the costumes recommendation module, which is mainly composed of knowledge base, inference engine, and the explanation mechanism. There are two parts in the knowledge base. Namely, the fact base and the rule base. The fact base stores on sell-costumes information while the rule base stores expert knowledge about costume matching. The expert knowledge is generated based on an algorithm named production rules. According to theory of inference engine, the CERS could realize the personalized costumes recommendation based customer's physical features, and present the final recommendation list to the customer. The recommendation result is also explained to the customer with the Explanation mechanism.

This paper mainly focuses on the design and implementation of classification of physical features. And we take the classification of skin-color as an example to describe how to acquire recommended costumes in details.

3 Classification of Physical Features on Skin-Color

It is known that Ms. Carroll Jackson proposed a color theory of four seasons [5] in 1970, dividing the skin-color into four seasons of spring, summer, autumn, and winter. Spring and autumn skin-colors fit warmer colors, while summer and winter skin-colors fit cooler colors. Specifically, the spring-type people's skin-color is light ivory, warm beige, delicate texture and a sense of transparency, so spring-type people are suitable for light, soft color; summer-type people's skin-color is pink-white, milky, brown skin, so summer-type people are suitable for light blue-violet, not suitable for too bright and pure color; autumn-type people's skin-color is dark-orange, dark-beige, so autumn-type people are suitable for thick golden to promote mature and elegant temperament; winter-type people's skin-color is white and slightly olive without blush, so the most suitable color is pure, bright, high glossiness.

Therefore, we analyze the skin-color and color for matching. First, we process the customers' image to obtain their skin-color. Here the skin-color mainly refers to their face color. So the image should be processed to locate the face part. After that, HOG and SVM are applied to extract and classify skin-color of the face part [6]. Then based on the color theory of four seasons stored in the rule base of the Expert system, we are able to recommend suitable costume color with the customers' skin-colors.

3.1 Image Process

The step of process customers' images includes the extraction of RGB components, the conversion of color-space, the segmentation and expansion of skin-color regions, and the location of human face. First an image is uploaded and the RGB components are extracted respectively. Then the RGB is transferred into the YCbCr color-space. This color-space can identify the skin-color region according to the conversion formula. Second, the image is segmented and expanded and the human face is located.

In order to segment the face-region from the non-human face-region, a skin-color model is required to maintain high robustness under varied skin-colors and lighting conditions. The RGB color-space is not suitable for the skin-color model because the three primary colors (r, g, b) not only represent the chroma but also the brightness. So we use the YCbCr space in which the chroma and brightness are separated. It is beneficial to be less affected by changes in brightness, and it can better limit the skin-color distribution because of its independently two-dimensional distribution.

In this paper, the human face is segmented from backgrounds so as to reduce the influence of background-colors on skin-colors and reduce the amount of calculation. We use expansion to remove discrete information which causes disturbances to the location of human face. Additionally, we define a length-to-width ratio of the face as 0.8–1.6 according to a large number of image statistics. Regions that satisfy this will be outlined to lay a foundation for subsequent classification. Figure 2 shows one example of preprocess of image.

3.2 Extraction of Skin-Color with HOG Method

The HOG, which extracts features by calculating and counting histograms of gradient directions in local regions of the image is a feature descriptor used for object detection [7]. First, we divide the image into small connected regions called cell units; secondly,

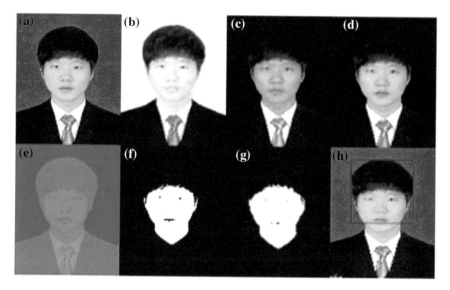

Fig. 2 Result of processing of image. **a** Original image; **b** R channel image; **c** G channel image; **d** B channel image; **e** image in YCbCr; **f** segmented image; **g** expanded image; **h** located image

we collect the gradient and directional histograms of pixel points in cell units; finally, we combine these histograms.

3.3 Classification of Skin-Color with SVM Algorithms

The basic idea of SVM is to find an optimal linear plane by solving a convex optimization problem. For the classification problem, datasets cannot be processed in the low-dimensional space but in high-dimensional space via linearization. We employ Mercer kernel theorem to increase the sample dimensions. In this paper, the SVM multi-classifier introduces a nonlinear mapping function φ. Fisher's linear discriminant criterion is selected in the high-dimensional space to optimize the SVM kernel function [8]. Here we select the kernel function of the mapping φ as the RBF kernel function:

$$K(\mathbf{x}_i, \mathbf{x}_j) = \exp(-\gamma^* \cdot ||\mathbf{x}_i - \mathbf{x}_j||^2), \qquad (1)$$

where γ^* is the optimal parameter determined by the Fisher criterion; x_i and x_j are skin-color feature vectors respectively.

Some practical classification problems are converted into multiple binary classifications via "one-to-one" or "one-to-many". We adopt the method of "one-to-many". Namely, some samples are classified into one class, and the remaining samples are

Table 1 The estimated result of SVM classifiers

Project	Training sets	Test sets	Accuracy of training (%)	Accuracy of test (%)
Spring	217	68	93.75	89.87
Summer	217	68	92.37	94.61
Autumn	217	68	83.28	81.37
Winter	217	68	96.73	91.28

classified into another class. According to the color theory of the four seasons, there are four types of skin-color, which are stipulated as A, B, C, and D in order. Therefore, when the skin-colors are extracted, they are marked as

(1) Class A is positive, (B, C, D) is negative; (2) Class B is positive, (A, C, D) is negative; (3) Class C is positive, (A, B, D) is negative; (4) Class D is positive, (A, B, C) is negative.

Here, the first classifier assumes that the "spring" sets are a positive class, then the remaining seasons sets are considered negative classes. A classifier decision function is constructed:

$$f_i^1(\mathbf{x}) = \text{sgn}(w_i\mathbf{x} + b_i), \tag{2}$$

where w_i, and b_i are parameters, and x is sample. Use the same method to train other classifiers and construct decision functions f_i^2, f_i^3, f_i^4 respectively. Ideally, one input sample may only respond one positive value among the four classifiers, which makes it easy to determine the type of skin-color. For example, when entering a sample into four classifiers in order, only the value of f_i^4 is positive, and the other three decision functions are negative and lower than the former, so the sample is belong to "winter". This paper selects images from the LFW face image database to make cross validation. The image datasets include 868 training sets and 272 test sets, namely, 80% are training sets and 20% are test sets. Then, we estimate the classifier with test sets, and if the accuracy is higher than 60%, the classifier is available. The estimated result is shown in Table 1 and the classified result is shown in Fig. 3.

4 Costumes Recommendation Module

The costumes recommendation module is mainly composed of an Expert System which includes the knowledge base, the inference engine, and the explanation mechanism. It is known that inference engine is the core of the expert system. The inference engine takes forward inference [9] to simulate the thinking process of the costumes expert, and perform specific target search and query according to inference rules. In this paper, the inference engine adopts the blackboard model, which adopts the

Fig. 3 Classified result of SVM classifiers. **a** Spring-type, accuracy of test: 87.68%; **b** Summer-type, accuracy of test: 90.21%; **c** Autumn-type, accuracy of test: 79.21%; **d** Winter-type, accuracy of test: 87.24%

forward inference approach, namely, starting from the fact base, inferring the results of the rules according to the inference strategy or expert knowledge.

The inference engine continuously gets unknown conclusions from known information. The costume recommendation system invokes the inference engine and constantly draws the conclusion from known information according to costume matching knowledge. The matching knowledge is mainly based on the principle of costume-color and the costume style knowledge. For instance, face-shape and shoulder-shape determine the suitable collar-shape of the clothing and the body-shape determines the suitable costume style. The proposed system can search the suitable clothe collocation for the customer on the basis of customer's skin-color and the body figure. Moreover, it stores the intermediate results into the knowledge base for further inference, and the unknown state of a problem is converted to a known state. Finally, the system searches the fact base for the on sell costume information in accordance with the results of the inference engine. The system generates the final clothe recommendation results and gives the explanation through the explanation mechanism. For the detail information on costumes recommendation expert system, please refer to another publication of our research group [10].

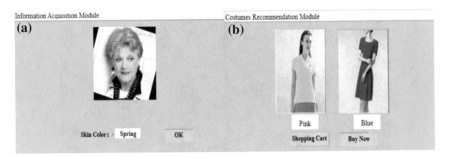

Fig. 4 Experimental results. **a** Customers' information acquisition; **b** Recommended results

5 Experimental Results

Suppose a customer uploads an image of his/her front face. After the upload is completed, the system firstly preprocesses the image; secondly, it accomplishes features extraction based HOG and classification based SVM respectively. Through the information acquisition module we can obtain the customer's physical features on skin-color as shown in Fig. 4a. So the skin-color is spring-type, the suitable costumes color will be pink and blue.

Considering other physical features of the customer, say, lean body-shape, inverted triangle face-shape, and narrow shoulder-shape, the recommendation conclusions are set as follows: the suitable costume-colors are pink and blue; the recommended costume styles are slim and routine; the collar-shape might be round, square, standing, high, polo, and horizontal. Then the recommendation system searches the on sell clothes information in the database to match the above conclusion, and the recommended results shown in Fig. 4b.

6 Conclusions

In this paper we mainly integrate image preprocessing, HOG feature extraction, SVM classifier, and Expert System to develop a physical features-based CERS. The proposed system could provide customers with intelligent and professional costumes recommendation services. Particularity, the customer's face color identification is illustrated in details. In the image preprocessing process, we obtain an accurate human face and frame it out by skin-color segmentation as well as facial ratio of the length and width. Then we achieve the extraction of skin-color feature with HOG and SVM algorithms. Finally, Expert System based on blackboard model is developed in the system. Experimental result shows how the proposed CERS system recommends suitable costume in accordance with the customer's physical features.

References

1. Mundhe, M., Phadnis, S., Rathi, P., Aghav, J.: A beacon based retail industry analytics framework. Int. J. Web Sci. Eng. Smart Devices **4**(1), 27–32 (2017)
2. Yang, E.K., Kim, S.-J.: 3D character virtual costume making software usability assessment—focusing on poser 3D character virtual costume making. J. Digital Des. **14**(1), 863–876 (2014)
3. Zheng A.: Research on apparel type recommendation method based on BP neural network, pp. 1–20. Cloth Institute, Zhejiang University of Science and Technology, Zhejiang (2010)
4. Kim, S., Han, D.S.: Real time traffic light detection algorithm based on color map and multilayer HOG-SVM. J. Broadcast Eng. **22**(1), 62–69 (2017)
5. Ibrahim, A.W., Sartep, H.: Grayscale image coloring by using YCbCr and HSV color spaces. Int. J. Mod. Trends Eng. Res. **4**(4), 130–136 (2017)
6. Agrawal, N., Bohra, S.: Face feature based extraction and classification for gender recognition and age estimation using fuzzy model: a review. Int. J. Recent Trends Eng. Res. **3**(4), 146–148 (2017)
7. Liu, Y., Zeng, L., Huang, Y.: An efficient HOG–ALBP feature for pedestrian detection. SIViP **8**(S1), 125–134 (2014)
8. Sunitha, B., Seetha, M.: Enhancing the performance of SVM classification based on fisher linear discriminant. Int. J. Res. Eng. Technol. **03**(17), 48–54 (2014)
9. Zamsuri, A., Syafitri, W., Sadar, M.: Web based cattle disease expert system diagnosis with forward chaining method. IOP Conf. Ser.: Earth Environ. Sci. **97**, 012046 (2017)
10. Mao, Q., Dong, A., Miao, Q., Pan, L.: Intelligent costume recommendation system based on expert system. J. Shanghai Jiaotong Univ. (Science) **23**(2), 227–234 (2018)

Cognitive Characteristics Based Autonomous Development of Clothing Style

Jiyun Li and Xiaodong Zhong

Abstract Due to the subjective characteristics and rapid change of fashion style, it is relatively hard to predefine the style feature in style classification systems. In this paper we present a cognitive characteristics based clothing style autonomous development model. By the addition of special domain related information to the classic itti visual attention model, we achieve the multi-object attention model of the clothing style. And based on this we implemented the autonomous development of clothing style recognition by Multi-Layer In-place Learning Network (MILN in short). Experiments prove the feasibility and effectiveness of our model.

Keywords Clothing style · Multi-object MILN · LCA · Autonomous development Visual attention

1 Introduction

As a major medium for communication between designers and consumers [1] style recognition and classification plays an important role in intelligent fashion design systems, in which different parts of the clothing can be chosen automatically to combine into a totally new design according to certain style definition. Yet unlike in architectural design where style can be seen as a collection of some basic relatively static objective style features [2], style definition in fashion design can be both subjective and time changing. In fashion design area, the usual style recognition practice is that experts establish correspondence between the style definition and features of the style by manually analyzing a large collection of designs that belong to a certain style and then the style of a given design is decided by extracting the features of the

J. Li (✉) · X. Zhong
School of Computer Sciences and Technology,
Donghua University, Shanghai 201620, China
e-mail: jyli@dhu.edu.cn

X. Zhong
e-mail: xdz@mail.dhu.edu.cn

© Springer Nature Switzerland AG 2019
W. K. Wong (ed.), *Artificial Intelligence on Fashion and Textiles*,
Advances in Intelligent Systems and Computing 849,
https://doi.org/10.1007/978-3-319-99695-0_11

design and matching the features the design owns with the features the style owns [3]. The style recognition method above demands the predefined mapping relationships between style definition and design features, thus it is only applicable to existing clothing style analysis and is really time-consuming. Due to the limitation of finite expressions in features, the kept on emerging new designs and style features which are inevitable in intelligent fashion design or recommendation systems cannot be covered by this method. Yet human beings can handle this easily. Originally people do not have a concept about the style and its features of designs, but the concept of a style and its associated features gradually begin to form and improve in the person's mind with the increasing of various design contact. This cognitive characteristic of human beings' can be modeled by Autonomous Mental Development (AMD). In AMD, machines' cognitions can grow through the interaction with the environment, thus it is not limited to the initial knowledge representation. Therefore this paper establishes a model by learning from the cognitive process of human style formation, classification, and feature extraction.

The paper is organized as follows. After a brief introduction of the research background and related works, we first presented the model framework. Visual attention model of clothing style is discussed in detail in Sect. 3; The autonomous development of clothing styles based on Multi-object MILN is given out in Sect. 4; Experiments design and discussion is given out in Sect. 5 and conclusions and further research directions are given out in Sect. 6.

2   Model Framework

The model framework of the autonomous development in clothing style is shown in Fig. 1. Where, the database for expert style stores the experts' style definition which will be used as a default setting during system initialization. The database for style feature stores the corresponding relationship between stored style and the features that the style has. The <T, S, P> database stores the corresponding relationships between the scene S and the location P of target objects T in the scene. We will obtain the location of clothing parts of the clothing image input by users through the visual attention module which simulates the human visual process.

MILN (Multi-Layer In-place Learning Network), which was proposed by Professor Weng Juyang [4], is an Artificial neural network that is based on the thought of Autonomous mental development. In MILN, Probability density of most samples will be concentrated on a certain direction which is called a lobe (Lobe). Here a modified Multi-object MILN is used to tackle the multi-object characteristics of fashion style and the module of extracting style features based on Lobe is used to analyze and extract the style features of clothing.

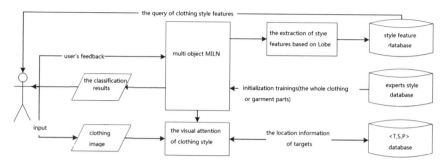

Fig. 1 Model framework

3 Visual Attention Model of Clothing Style

When we begin to analyze clothing style, we should pay attention to not only the clothing as a whole but also the parts which make the clothing as whole by combination. This model discussed below pays visual attention to clothing parts. Traditional visual attention models only mimic the Bottom-up visual attention of human visual system. They only focus on low level features of the design image, so it is difficult to reflect the high-level subjective semantic features of clothing parts. Although some improved attention models, such as itti visual attention model have incorporate Top-down channel into Bottom-up visual attention model [5], it is still not applicable in fashion style computation due to the strong area correlation existing in clothing parts.

In this paper we use MILN to learn the area related knowledge and then we add the channel of Top-down visual attention that is based on MILN to the traditional Bottom-up based visual attention model. And the final saliency map of a visual attention task is achieved by synthesizing the saliency maps of both the Bottom-up channel and the Top-down channel.

There are two channels, namely the channels of What and Where, in the Top-down visual attention channel based on MILN (Eysenck 2009). The What channel identifies the target objects, while the Where channel locates the target object.

4 The Autonomous Development of Clothing Styles Based on Multi-object MILN

4.1 The Autonomous Development Network of Multi-object MILN

The autonomous development network of multi-object MILN is an adaptation of the autonomous development network of MILN, in which the first layer of the network

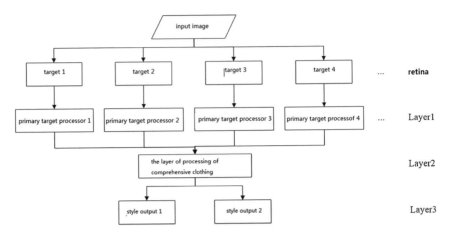

Fig. 2 The structure of multi-object autonomous development network of MILN

is divided into *n* independent processing units. Then the *n* different units begin to deal with different targets. The second layer synthesizes the result of the first layer comprehensively. The final results are output by the third layer.

4.2 The Autonomous Development of Clothing Style

On the basis of simple classification of MILN we construct our scene-based multi-object autonomous development network of MILN. The structure is shown in Fig. 2. This is a three-layer model.

Suppose an input image I consists of *n* target objects, T1, T2, T3, ..., T*n* that we can get from the input image through visual attention model discussed in Sect. 2. Each of these objects is projected on the retina as the input of the receptive field of the network. One input is processed by multi-receptive fields. After a preliminary process in layer 1 the target objects now g1, g2, ..., g*n*, respectively will be transferred to comprehensive clothing style processing layer, i.e., layer 2. Finally the clothing style results o1, o2, ..., o*m* output in layer 3.

5 Experiments and Discussions

We select 450 pieces from classic style and 450 pieces from elegant style as identification objectives. We take 200 pieces as preliminary training samples and make its number increase gradually. The numbers of samples were 220, 240, 260, 280, 300, 350, respectively. The numbers of training and testing processes were 10. We take the average recognition rate as our final output. The results of using single-object

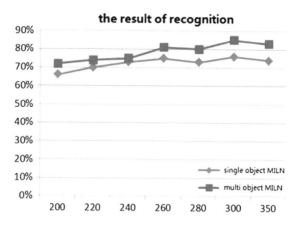

Fig. 3 The comparison of the recognition results of single-object and multi-object MILN

MILN and multi-object MILN are shown in Fig. 3. X axis is the number of training samples and Y coordinates is the precision rate of the recognition.

From Fig. 3 we can see that

(1) In the case of few training samples, the recognition results reflect more of the style definition from a statistics point of view. With the increment of interaction, i.e., by increasing the number of training samples, MILN gradually evolves and the style recognition results gradually approaches to the goal setting. The recognition result begins to stabilize when the number of training samples reaches about 300.

(2) Although the single-object MILN can develop the concept of clothing style, it has limitation for only taking the clothing as a whole thus with only one visual focus. The clothing style generally is made from a combination of clothing parts styles. What's more we cannot deal with single visual focus effectively when we analyze styles of clothing, because there are dispersed multi-objectives in this view. We cannot get the whole features about styles of clothing types through the analysis for Lobe and it is difficult to get the types features of each clothing part. On the contrary, the input samples of multi-object MILN synthesize the targets which can decide clothing styles including not only the whole clothing but also clothing parts such as collars, sleeves and clothing pieces. Thus the recognition rate of multi-object MILN is 10% higher than that of single-object MILN in average.

6   Conclusions and Further Directions

Multi-object MILN is an adaptation of single object MILN by incorporation of the multi-object visual attention characteristics of human cognition into itself. By developing a style concept from different objects in an image then synthesizing the separate results together it mimic the clothing style formation and recognition process of human beings well. Experiments show that the recognition rate for clothing styles by multi-object MILN is more than 80%. Besides we can analyze the style features of the clothing parts by extracting style information from the lobe components of MILN due to multi-object characteristic, which can further guide the fashion design.

At present our model can only be used for the recognition and analysis of styles of the clothing drawn in line sketch. We will study the possibility of using this model for natural fashion images later on. In the mean time, improving the efficiency and accuracy rate of the recognition algorithm will be a long-term research direction in the future.

References

1. Huang, Q., Sun, S.: The research progress for the calculation in product styles. J. Comput. Aided Des. Comput. Graph. **18**, 1629–1636 (2006)
2. Chan, C.-S.: Can style be measured. Des. Stud. **21**, 227–291 (2000)
3. Feng, L., Liu, X.: First exploration of the ways to quantify clothing styles. J. Donghua Univ. (Nat. Sci.) **30**(1), 57–61 (2004)
4. Weng, J., Luwang, T., Lu, H.: A multilayer in-place learning network for development of general invariances. Int. J. Humanoid Rob. **4**(2), 281–320 (2007)
5. Robert, J., Laurent, I.: Beyond Bottom-up: incorporating task-dependent influences into a computational model of spatial attention. In: IEEE Conference on Computer Vision and Pattern Recognition, pp. 1–8 (2007)
6. Huang, Q., Sun, S.: The recognition method for product styles based on matching features. China Mech. Eng. **14**, 1836–1838 (2003)
7. Weng, J., Luciw, M.: Dually optimal neuronal layers: lobe component analysis. IEEE Trans. Auton. Ment. Dev. **1**(1), 68–85 (2009)
8. M, W., M, T.: Cognitive Psychology: A Student's Handbook, 6th ed., Psychology Press (2010)
9. Chen, J.: Researches for the mode of style operation which is used to design product style. Ind. Des. **28**, 111–115 (2000)
10. Cheng, G., Li, J.: Research on clothing style preference model based on interactive genetic algorithm. Comput. Appl. Softw. **28**(2), 229–231 (2011)

Fabric Identification Using Convolutional Neural Network

Xin Wang, Ge Wu and Yueqi Zhong

Abstract Image-based fabric retrieval technique can help to develop new fabrics and manage products. Efficiently extracting features from fabric images is the key to enhance the practicality of this technology. In this paper, convolutional neural network is trained with a dataset of 19,894 different yarn-dyed fabric patterns. Center loss architecture is added to further improve the discriminative power of the network. By properly sampling from original images, the network model can efficiently extract discriminative features and achieve a retrieval accuracy of 99.89% on our test set. This performance maintains well when simpler deep architecture is used, but decreases quickly if the contents of fed fabric image are reduced.

Keywords Fabric retrieval · Convolutional neural network · Similarity learning

1 Introduction

Fabrics have varied and diverse patterns. However, the industry does not have an efficient mechanism for managing numerous fabrics to avoid developing the already existing products or designing similar products [1]. Retrieving fabric based on image contents can well satisfy this need. It can quickly find fabrics with identical or similar visual effect from a database, which facilitates the industry management. The premise of this technique is to encode fabric images to features which are discriminating between different fabrics and similar between identical fabrics. In this study, convolutional neural network is trained to extract features from images with a dataset containing 19,894 kinds of yarn-dyed fabric patterns. To improve the discriminative power of the deeply learned features, center loss [2] is introduced to total loss func-

X. Wang · G. Wu · Y. Zhong (✉)
Donghua University, Shanghai 201620, China
e-mail: zhyq@dhu.edu.cn

G. Wu · Y. Zhong
Key Lab of Textile Science and Technology, Ministry of Education,
Shanghai 201620, China

© Springer Nature Switzerland AG 2019
W. K. Wong (ed.), *Artificial Intelligence on Fashion and Textiles*,
Advances in Intelligent Systems and Computing 849,
https://doi.org/10.1007/978-3-319-99695-0_12

tion. To train a well-performed network, different sampling areas of original images are tested. The result shows that larger sampling area leads to a better performance. To summarize, the main contributions of this study include:

- A convolutional neural network with center loss architecture is introduced to fabric retrieval problem and experiment on a dataset with more than 10,000 patterns.
- By experimenting on the dataset, the effects of sampling area from original images and complexity of models are detailed analyzed.

2   Related Works

Various algorithms have been proposed to encode fabric images to feature vectors. Nanik [3] used both fractal-based texture features and HSV features to represent fabrics. Jing [4] proposed a method combining color moments and gist features to classify 700 fabrics images. Cao [5] proposed a fabric image matching algorithm by constructing the bag of visual words using K-Means algorithm. However, a common problem of these methods is they were evaluated on a dataset of limited size, and their retrieval performance needed be further improved.

Convolutional neural network and deep learning have made a great breakthrough in past five years, especially on the tasks related to computer vision. In the field of face recognition, it achieves 99.77% in Labeled Faces in the Wild (LFW) dataset,[1] which is the state-of-the-art performance and beyond human performance 99.20%. For recognition problems, several ways can be used to enhance model discriminative power like contrastive loss [6], triplet loss [7] and center loss [2], etc. In this paper, a deep architecture combining softmax cross-entropy and center loss is used considering its competitive performance and time efficiency. This paper begins with dataset description. Then, network architecture and evaluation metric will be explained in more details. Finally, the performance and robustness of models will be further discussed.

3   Method

3.1   Dataset of Yarn-Dyed Fabric Patterns

Totally 19,894 yarn-dyed fabric images were prepared in our dataset. They are referred as original images. The resolution of original images is 1800×1800 pixels. Examples of original images are shown in Fig. 1a. Each original image was cropped and resized into two sub-images. The schemes of these preprocessing are listed in Table 1. Before feeding into models, sub-images are randomly flipped, cropped, and

---

[1]http://vis-www.cs.umass.edu/lfw/results.html.

(a)   (b)

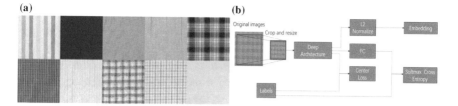

Fig. 1 **a** Examples of original images. **b** Network architecture combining softmax cross entropy and center loss

Table 1 Different scheme to preprocess original images

Original	Crop	Resize	Input	Label
1800 × 1800	299 × 299	299 × 299	224 × 224	C224
1800 × 1800	224 × 224	224 × 224	180 × 180	C180
1800 × 1800	180 × 180	180 × 180	160 × 160	C160
1800 × 1800	1600 × 1600	299 × 299	224 × 224	CR224
1800 × 1800	1600 × 1600	224 × 224	180 × 180	CR180
1800 × 1800	1600 × 1600	180 × 180	160 × 160	CR160

rotated. For the convenience of explanation, we use C224 to denote the preprocessing makes the final input of 224 × 224 pixels. C180 and C160 have the same meaning except for the input size. CR224 indicates the preprocessing where original images are cropped into 1600 × 1600 pixels followed by resizing to 224 × 224 pixels. The dataset is split into training, validation, and test set with a ratio of 6:2:2.

3.2 Network Architecture

The network architecture employed in this paper is illustrated in Fig. 1b. Images are transformed into feature vectors with a dimension of 128 called pre-logits. Another fully connected layer is added after pre-logits with output dimension equals to the categories of fabrics for training. Softmax cross-entropy can help classify patterns between different categories, which are formalized as

$$L_{\times} = \sum_{i=1}^{m} -\log\left(\frac{e^{X_i^{y_i}}}{\sum_j e^{X_i^j}}\right) = \sum_{i=1}^{m}\left(-X_i^{y_i} + \log\sum_j e^{X_i^j}\right) \tag{1}$$

m is the number of training samples, $X_i = f(x_i, w, b)$ is the predicted scores. Adding center loss could further improve the discriminative power, which is defined as

$$L_c = \frac{1}{m} \sum_{i=1}^{m} \|x_i - c_{y_i}\|^2 \qquad (2)$$

where x_i is the learned features from image, c_j is center of features of label j. Centers are initialized to zeros and can be updated with the progress of training. For each iteration, center of each class is updated like this, for each x_i:

$$c_{y_i}^{t+1} = c_{y_i}^{t} - (1 - \alpha)(c_{y_i}^{t} - x_i) \qquad (3)$$

where α is the factor controls update rate of centers. The definition of total loss is

$$L_{\text{total}} = L_\times + \lambda L_c \qquad (4)$$

Where λ is the weight of center loss to total loss. Two deep architectures are adopted i.e. Inception-resnet-v1 [8] and SqueezeNet [9]. RMSPROP optimizer is used for training with 80 epochs. Learning rate is initialized to 0.1, then decreases to 0.01 after 64 epochs and 0.001 after 77 epochs. Wight decay is set to $5e^{-5}$. Keep probability of each dropout layer is 0.8. Parameters of center loss are $1e^{-2}$ for λ, 0.9 for α. The TensorFlow implementation[2] was tested on an Intel i5-6500 @ 3.2 GHz with 8 GB memory and NVIDIA GTX 1080 GPU. Training an Inception-resnet-v1 model with CR224 preprocessing takes roughly 8 h. After training, it takes average 0.006 s to extracting features from each image, and takes average 0.03 s to retrieve one images from test sets containing 7978 images.

4 Evaluation

To evaluate the performance of the trained model, two metrics are used. The first one is top-n analysis. For each query image, if their identical fabric images appear in the top n most similar images, it is thought as correctly retrieved. Ratio of correctly retrieved images to all images is retrieving accuracy. Euclidean distance is used to measure the similarity between embedding vectors. Retrieval examples with CR224 preprocessing scheme are shown in Fig. 2. The second method is threshold judgment. Given a pair of images, a threshold of their embedding vector distance is used to judge whether they are identical or not. Four indices are used for evaluation, i.e., accuracy, VAL, EER, AUC. The accuracy of a specific threshold d is

$$\text{Accuracy} = \frac{|TA(d)| + |TR(d)|}{P_{\text{same}} + P_{\text{diff}}} \qquad (5)$$

where P_{same} is all image pairs of the same fabric, P_{diff} is all image pairs of different fabric. $TA(d)$ is the set of all true accepted same image pairs, $TR(d)$ is the set of

---

[2]More implementing details can be found at https://github.com/WangXin93/FabricID.

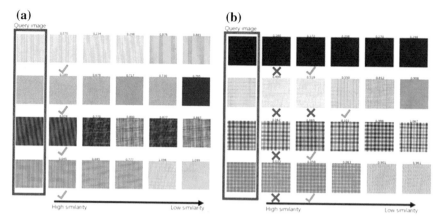

Fig. 2 **a** Correct search examples in top-1 analysis. **b** Incorrect search examples in top-1 analysis. The numbers above images are Euclidean distance between the retrieved image and the query image

all true rejected different image pairs. Another index VAL while FAR equals 0.001 is defined as

$$\text{VAL}(d) = \frac{|TA(d)|}{P_{\text{same}}}, \text{FAR}(d) = \frac{|FA(d)|}{P_{\text{diff}}}, \tag{6}$$

where FA is the false accepted set. Equal error rate (EER) is the common value of false accepted rate and false rejected rate when they are equal. Area under curve (AUC) is the area between ROC curve and x axis.

5 Results and Discussion

To have an intuitive overview of fabric embedding, t-SNE [10] was used to visualize image features in 3D space. After iterating 5000 steps with perplexity as 50, locations of these images are shown in Fig. 3a. It can be observed that images with similar appearance gather together which implies these features have a good representation of fabric images. Figure 3b shows top-n retrieval accuracy of the different preprocessing scheme. With crop area increasing, model performance improves. The best performance appears while using preprocessing scheme CR224. The top-1 accuracy reaches 99.89% and only has nine incorrect predictions, then in top-3 test, the accuracy becomes 100%. The incorrect retrieving examples are shown in Fig. 2b. When changing the architecture to SqueezeNet, the top-1 accuracy is still larger than 97.5%. However, the performance decreases significantly once using only cropping preprocessing. The results of threshold evaluation are shown in Table 2. Once the input size increases, the model performs relatively better, which is the same as top-n

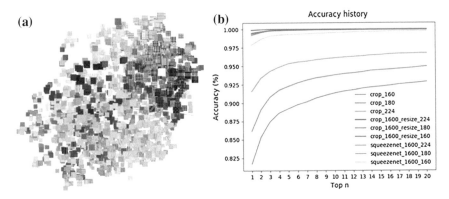

Fig. 3 **a** Visualization of fabric embedding using t-SNE. **b** Top-n retrieval accuracy, SCR224 has the same meaning as CR224 except for using SqueezeNet architecture

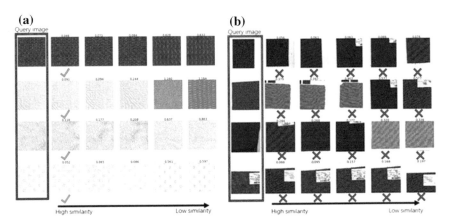

Fig. 4 **a** Correct search examples in top-1 analysis. **b** Incorrect search examples in top-20 analysis

analysis. It can be assumed that the effect of model complexity is less than image contents fed into the model in this experiment.

- To simulate the scenario where new types of fabric products are registered into the current database, new samples composed of 201 scanned fabrics were added into the test set. Their retrieval results are shown in Fig. 4. By analyzing all incorrect retrieval samples, what in common is they are either partially covered, or have a blank part in images. This result indicates the trained model mistakenly thinks attached labels and blank area as features of fabrics.

Table 2 Metric results with different preprocessing

Metric	C160	C180	C224	CR160	CR180	CR224	SCR160	SCR180	SCR224
Accuracy	0.969	0.980	0.985	0.999	1.000	1.000	0.998	0.999	0.999
VAL@1e-3	0.88200	0.91900	0.93998	0.99923	0.99975	1.00000	0.99775	0.99876	0.99898
AUC	0.991	0.995	0.996	1.000	1.000	1.000	1.000	1.000	1.000
EER	0.036	0.024	0.018	0.000	0.000	0.000	0.002	0.001	0.001

Labels have the same meaning with Fig. 3

6 Conclusion

According to the experiment results, convolutional neural network combining soft-max cross-entropy and center loss can effectively extract discriminative features from yarn-dyed fabric pattern images. The best result in our experiment had only nine errors in top-1 analysis by testing on 3978 kinds of patterns. Comparing different conditions, we found that sampling area from original fabric images had more effect than deep architecture complexity. However, these features are not robust while there are blank areas or attached labels on fabric images.

Acknowledgements This work is supported by National Natural Science Foundation of China (Grant No. 61572124).

References

1. Kuo, C.-F., Lee, C.-L., Shih, C.-Y.: Image database of printed fabric with repeating dot patterns part (I)—image archiving. Text. Res. J. **87**, 2089–2105 (2017). https://doi.org/10.1177/00405 17516663160
2. Wen, Y., Zhang, K., Li, Z., Qiao, Y.: A Discriminative Feature Learning Approach for Deep Face Recognition BT. In: Leibe, B., Matas, J., Sebe, N., Welling, M. (eds.) Computer Vision—ECCV 2016: 14th European Conference, Amsterdam, The Netherlands, 11–14 Oct. 2016, Proceedings, Part VII, pp. 499–515. Springer, Cham (2016)
3. Suciati, N., Herumurti, D., Wijaya, A.Y.: Fractal-based texture and HSV color features for fabric image retrieval. In: IEEE International Conference on Control System, Computing and Engineering, pp. 178–182 (2016)
4. Jing, J., Li, Q., Li, P., Zhang, L.: A new method of printed fabric image retrieval based on color moments and gist feature description (2016) https://doi.org/10.1177/0040517515606378
5. Cao, Y., Zhang, X., Ma, G., Sun, R., Dong, D.: SKL algorithm based fabric image matching and retrieval. In: International Conference on Digital Image Processing, p. 104201F (2017)
6. Chen, Y., Chen, Y., Wang, X., Tang, X.: Deep learning face representation by joint identification-verification. In: International Conference on Neural Information Processing Systems, pp. 1988–1996 (2014)
7. Schroff, F., Kalenichenko, D., Philbin, J.: FaceNet: A unified embedding for face recognition and clustering. In: Proceedings of the IEEE Conference on Computer Vision and Pattern Recognition, pp. 815–823, 07–12 June 2015. https://doi.org/10.1109/cvpr.2015.7298682
8. Szegedy, C., Ioffe, S., Vanhoucke, V., Alemi, A.: Inception-v4, Inception-ResNet and the Impact of Residual Connections on Learning (2016)
9. Iandola, F.N., Han, S., Moskewicz, M.W., Ashraf, K., Dally, W.J., Keutzer, K.: SqueezeNet: AlexNet-level accuracy with 50x fewer parameters and <0.5 MB model size, pp. 1–13 (2016). https://doi.org/10.1007/978-3-319-24553-9
10. Maaten, L.V.D., Hinton, G.: Visualizing data using t-SNE. J. Mach. Learn. Res. **1**(620), 267–284 (2008). https://doi.org/10.1007/s10479-011-0841-3

Discrete Hashing Based Supervised Matrix Factorization for Cross-Modal Retrieval

Baodong Tang, Xiaozhao Fang, Shaohua Teng, Wei Zhang and Peipei Kang

Abstract Cross-modal hashing is a method which projects heterogeneous multimedia data into a common low-dimensional latent space. Many methods based on hash codes try to keep the relationship between text and corresponding image, and relax the original discrete learning problem into a continuous learning problem. However, these methods may produce ineffective hash codes since they do not make full use of the relationship between different modalities and simply relax the discrete binary constraint into a continuous problem. Collective matrix factorization (CMF) has achieved impressive results in mining semantic concepts or latent topics from image/text. In this paper, we propose a new supervised learning framework which unifies CMF method that maximizes the correlation between two modalities and discrete cyclic coordinate descent (DCC) method that solves NP-hard problems, which ensures that the hash codes generated in the cross-modal are more accurate and efficient. Experiments on three benchmark data sets show the effectiveness of the proposed method.

Keywords Collective matrix factorization · Discrete cyclic coordinate descent
Cross-modal retrieval

B. Tang · X. Fang (✉) · S. Teng · W. Zhang · P. Kang
School of Computer Science and Technology,
Guangdong University of Technology, Guangzhou, China
e-mail: xzhfang168@126.com

B. Tang
e-mail: bdtanggdut@126.com

S. Teng
e-mail: shteng@gdut.edu.cn

W. Zhang
e-mail: weizhang@gdut.edu.cn

P. Kang
e-mail: ppkanggdut@126.com

© Springer Nature Switzerland AG 2019
W. K. Wong (ed.), *Artificial Intelligence on Fashion and Textiles*,
Advances in Intelligent Systems and Computing 849,
https://doi.org/10.1007/978-3-319-99695-0_13

1 Introduction

With the high pace of technology, the multimedia data grow rapidly, e.g., a blog with text, image and video, the Weibo news with pictures and associated texts, WeChat circle of friends and so on. There have many methods about retrieval in the community of computer vision, machine learning, and pattern recognition. The methods that generate a unique binary sequence of an image or video are more and more attractive, such as generating a unique binary sequence for each image [1], termed as hash coding. The bit *xor* operation is applied when calculating hamming distance between hash codes [2]. This makes it possible and efficient to use hash codes for modality-specific or cross-modal retrieval. There are two kinds of cross-modal hashing retrievals. The first one is to learn hash codes for each mode and retrieve under the corresponding mode. The second is to learn unified hash codes for multi-modal and use it to perform modality-specific or cross-modal retrieval. For example, in practical applications, we may encounter the necessity of pictures retrieval based on a word or a sentence.

Generally, cross-modal hashing can be divided into unsupervised [1], semi-supervised [3], and supervised [5] methods. The key to distinguish the supervised and unsupervised is whether they can take advantage of the label information in the methods and embody the information in hash codes. Supervised methods extract useful semantic information from the class labels of data sets. Usually, supervised cross-modal hashing approaches can achieve better retrieval performance.

The goal of supervised cross-modal hashing is to maximize the correlation within the category, and weaken the correlation between categories. Many related studies are based on latent subspace learning. For example, Canonical Correlation Analysis (CCA) [4] is a popular method that learns a common latent subspace from multi-modal. Collective matrix factorization (CMF) [8] is also a latent subspace learning method which can simultaneously decompose multi-modal data matrices, and each matrix represents the relationship between two modals or even more.

Motivated by the promising results of CMFH [6] and SDH [7] in single-modal and cross-modal retrieval, we propose a new approach termed as discrete hashing based supervised matrix factorization (SMFDH). It is a supervised learning framework which unifies CMF that maximizes the correlation between two modalities and discrete cyclic coordinate descent (DCC) that solves mixed-integer optimization problems(NP-hard). It learns unified hash codes from multi-modal data produced by the modality-specific hash functions. Through decomposition sample matrix we can obtain the common factor V (the common factor in two reconstructions). With discrete constraints on hash codes, SMFDH can improve the accuracy of retrieval. SMFDH approach can be extended to multi-modalities, but in this paper, we mainly focus the problem of cross-modal retrieval (image and text). Figure 1 shows the pipeline of SMFDH.

The remainder of this paper is organized as follows. We will introduce the details of the proposed method in Sect. 2. Section 3 shows the experiments and analysis. Conclusions are summarized in Sect. 4.

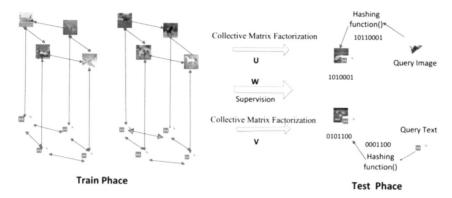

Fig. 1 The pipeline of SMFDH. In the train phase, the original data are projected to the latent semantic space, and acquire the projection matrix U and V, and correlation matrix W is obtained under the supervised learning framework to learn an unified hash binary code. In the test phase, through the modality-specific hash function to produced the hash codes

2 Proposed Method

2.1 Smfdh

We assume $X = \{x_i\}^n$ has n samples, $X^{(1)}$ is the image feature matrix and $X^{(2)}$ is the text feature matrix, Y is class labels of the samples. We aim to learn a matrix $B = \{b_i\}_{i=1}^n \in \{-1, 1\}^{L \times n}$ to preserve the semantic similarities for data set. L is the length of hash codes, b_i is the hash codes for x_i. Without of generality, we assume that the matrix $X^{(1)}$ and $X^{(2)}$ are all zero-centered, i.e., $\sum_{i=1}^n x_i^{(1)} = 0$ and $\sum_{i=1}^n x_i^{(2)} = 0$.

The multi-modal data with different labels should have discriminative hash codes, while those with the same label should have similar binary codes. We assume that every kind of image or text will generate its unique binary code sequence, and the high-quality binary sequence can efficiently classify the original data set.

$$y = G(b) = W^T b = \left[w_1^T b, \dots, w_c^T b\right]^T, \tag{1}$$

where $w_k \in R^{L \times 1}$ ($k = 1, 2, \dots, c$) is the classification vector for class k. y is the label vector that indicates which class of x belongs to.

We have the following optimization formula:

$$\min_{B,U,V,W} \sum_{i=1}^n \left\|Y - W^T B\right\|_F^2 + \lambda \|W\|_F^2 + \alpha_1 \left\|X^{(1)} - UB\right\|_F^2 + \alpha_2 \left\|X^{(2)} - VB\right\|_F^2 \tag{2}$$

$$\text{s.t.} B \in \{-1, 1\}^{L \times N},$$

where λ is a penalty parameter, α_1 and α_2 are the balance parameters. $\|.\|_F^2$ is the Frobenius norm for matrices. The first term expects discriminative binary codes for supervised classification. The last two terms represent that the original image and text feature matrices can be decomposed into a common low-dimensional hamming space by introducing two different matrices U and V.

2.2 Optimization Algorithm

The optimization problem of (2) is non-convex for each variable, but we can iteratively update each variable by fixing others.

Step1: (update U,V): For problem (2), by fixing B and W, it can be transformed into:

$$\min_{U} \sum_{i=1}^{n} \alpha_1 \left\| X^{(1)} - UB \right\|_F^2 \tag{3}$$

$$\min_{V} \sum_{i=1}^{n} \alpha_2 \left\| X^{(2)} - VB \right\|_F^2. \tag{4}$$

We can easily get $U = \left(BB^T \right)^{-1} X^{(1)} B^T$ and $V = \left(BB^T \right)^{-1} X^{(2)} B^T$.

Step2: (update W): If we fix U, V, and B

$$W = \left(BB^T + \lambda I \right)^{-1} BY^T. \tag{5}$$

Step3: (update B):
When fix other variables, we rewrite problem (2) as

$$\min_{B} \left\| Y - W^T B \right\|_F^2 + \alpha_1 \left\| X^{(1)} - UB \right\|_F^2 + \alpha_2 \left\| X^{(2)} - VB \right\|_F^2 \text{ s.t } B \in \{-1, 1\}^{L \times N} \tag{6}$$

Inspired by the DCC algorithm [7], we can solve one row of B by fixing all other rows, which is also called *bit*-by-*bit*. We can rewrite (6) as follows:

$$\min_{B} \| Y \|_F^2 + 2Tr\left(Y^T W^T B \right) + \left\| W^T B \right\|_F^2 + \alpha_1 \left[\left\| X^{(1)} \right\|_F^2 - 2Tr\left(X^{(1)T} UB \right) + \| UB \|_F^2 \right]$$
$$+ \alpha_2 \left[\left\| X^{(2)} \right\|_F^2 - 2Tr\left(X^{(2)T} VB \right) + \| \| VB \| \|_F^2 \right] \tag{7}$$

which is equivalent to

$$\min_{B} \sum_{i=1}^{n} \left\| W^T B \right\|_F^2 - 2Tr\left[Y^T W^T + \alpha_1 \left(X^{(1)T} UB \right) + \alpha_2 \left(X^{(2)T} VB \right) \right]. \tag{8}$$

Define

$$Q = WY + \alpha_1 X^{(1)^T} U + \alpha_2 X^{(2)^T} V \tag{9}$$

is equal to

$$\min_{B} \sum_{i=1}^{n} \left\| W^T B \right\|_F^2 + 2Tr\left(Q^T B\right) \tag{10}$$

where $Tr(.)$ is the trace of matrix. Let q^T is the rth row of Q, and Q' is the matrix of Q that excludes q. z^T be the rth row of B, and B' is the matrix of B excluding z. Similarly, e^T the rth row of W, and W' the matrix of W excluding e.

We can rewrite problem (10) as

$$\min_{Z} \left(z e^T W'^T B' - q^T \right) z. \tag{11}$$

This problem has the optimal solution

$$z = \text{sgn}\left[q - B'W'e\right]. \tag{12}$$

sgn(.) is the sign function, which outputs $+1$ for positive numbers and -1 otherwise. In the phase of cross-modal retrieval, such as retrieving the corresponding text with image or vice versa, the modality-specific hash function will produce hash codes by $b_i = U^{-1}X^{(1)}$ or $b_i = V^{-1}X^{(2)}$. The optimization of SMFDH is summarized in Algorithm 1.

Input: Image feature matrix $X^{(1)}$ and text feature matrix $X^{(2)}$; code length L;parameters $\lambda,\alpha_1,\alpha_2$;.

Initialization: W, U, V by random matrices respectively.

1. Center $X^{(1)}$ and $X^{(2)}$ by means.

2. Repeat

3. Fix other variables and solve (5) to update W;

4. Fix other variables and solve (3) (4) to update U, V,

5. Learn B bit by bit according to (12);.

6. Until the objective (2) converges.

Output: The hash codes B and matrices U and V.

---

3 Experiment

To verify the effectiveness of SMFDH, we conduct experiments on three popular data sets, i.e., Wiki [9], MIR-Flickr [10] and NUS-WIDE [11] which are widely used in cross-modal retrieval methods. Each of them has texts and the corresponding image. Finally, we will analyze the result and efficiency of the SMFDH.

3.1 Experimental Setup

Database description. We use the most popular three data sets for testing. Wiki MIR-Flicker and NUS-WIDE.

Table 1 is the configuration of our experiments, which includes the data sets size, training set size, test set size, and the number of categories.

We perform two kinds of cross-modal retrieval tasks. "Txt2Img" uses text queries to search corresponding images and "Img2Txt" uses image queries to search relevant texts.

A widely indicator *mean average precision* (mAP) is used in the experiments [8–15]. The definition of mAP is as follows:

$$\text{mAP} = \frac{1}{N} \sum_{j=1}^{n} \frac{1}{T} \sum_{i=1}^{l} \text{Precision}(i)\delta(i). \tag{13}$$

We assume that it is relevant as long as there is at least one label correlation. Precision(i) represents the precision of top i retrieved entities, $\delta(i) = 1$ indicates ith instance is relevant to the query, $\delta(i) = 0$ otherwise. T is the number of ground-truth relevant instances in the database for the ith query, and N is the number of the instances in query set. The mAP is the average AP of all queries.

Table 1 Genreal statistics of three data sets

Data sets	Wiki	MIR-Flicker	NUS-WIDE
Data set size	2866	15902	186577
Training set size	2173	15902	5000
Test set size	693	863	1866
Number of categories	10	24	10

Table 2 The performance and baselines on three data sets with various code lengths

Methods/Data sets		Wiki				MRI-Flicker				NUS-WIDE			
		16bits	32bits	64bits	128bits	16bits	32bits	64bits	128bits	16bits	32bits	64bits	128bits
Img2 Txt	LSSH	0.2083	0.2209	0.2214	0.2207	0.5748	0.5804	0.5797	0.5816	0.3955	0.3975	0.4009	0.4018
	CMFH	0.2074	0.2252	0.2357	0.2414	0.5861	0.5835	0.5844	0.5849	0.4327	0.4284	0.4257	0.4236
	CRH	0.1975	0.1960	0.1978	0.1939	0.5826	0.5745	0.5726	0.5718	0.4536	0.4579	0.4696	0.4713
	SCM	0.2150	0.2330	0.2437	0.2591	0.6069	0.6324	0.6435	0.6475	0.4910	0.5005	0.5006	0.5029
	SMFDH	0.3189	0.3127	0.3215	0.3314	0.6237	0.6362	0.6467	0.6487	0.5321	0.5341	0.5412	0.5440
Txt2 Img	LSSH	0.5021	0.5208	0.5277	0.5325	0.5798	0.5727	0.5732	0.5722	0.4363	0.4299	0.4303	0.4233
	CMFH	0.4874	0.5117	0.5253	0.5354	0.5937	0.5919	0.5931	0.5919	0.4710	0.4611	0.4578	0.4551
	CRH	0.2629	0.2614	0.2623	0.2615	0.5944	0.5913	0.5838	0.5811	0.4673	0.4714	0.4733	0.4777
	SCM	0.2130	0.2359	0.2472	0.2563	0.6121	0.6190	0.6276	0.6314	0.4618	0.4673	0.4690	0.4709
	SMFDH	0.5722	0.5716	0.5721	0.5739	0.6110	0.6123	0.6180	0.6156	0.5043	0.5101	0.5225	0.5231

3.2 Experimental Results and Analysis

We compare SMFDH with many of the existing algorithms. LSSH [12] and CMFH [13] learn unified binary codes for each training instance. CRH [14] and SCM [15] generate different binary codes for each modality of all instances and they are supervised methods to learn hash codes. The baseline and parameters of the algorithms in experiments follow the suggestions in their papers, and we choose the best results for them. The experimental results are shown in the Table 2. The code was run on Intel(R) core(TM) i5-6500 CPU @ 3.20 GHz RAM 8.0 GB.

From the results in Table 1, we can draw the following conclusions.

(1) The performance of our algorithm is obviously better than most of the methods in terms of mAP, especially on the databases of Wiki and NUS-WIDE.
(2) With the increase of hash code length, the mAP result becomes better. The reason may be that the longer binary code is, the more information they can retain, which can promote the recognition rate.

4 Conclusion

In this paper, we have proposed an effective cross-modal hashing approach termed as discrete hashing based supervised matrix factorization (SMFDH). It embeds heterogeneous multimedia data into a common low-dimensional hamming space to generate an unified hash codes for retrieval. We use efficient discrete optimization methods, making the quantization error as small as possible. The results on three benchmark data sets show the effectiveness of the proposed method.

Acknowledgements This work was supported in part by the Natural Science Foundation of China under Grant 61772141, in part by the Guangdong Provincial Natural Science Foundation, under Grant 17ZK0422, and in part by the Guangzhou Science and Technology Planning Project under Grants 201804010347.

References

1. Liu, W., Kumar, S., Kumar, S., et al.: Discrete graph hashing. In: International Conference on Neural Information Processing Systems. pp. 3419–3427 (2014)
2. Lin, Z., Ding, G., Hu, M., et al.: Semantics-preserving hashing for cross-view retrieval. 3864–3872 (2015)
3. Wang, J., Li, G., Pan, P., et al.: Semi-supervised semantic factorization hashing for fast cross-modal retrieval. Multimedia Tools Appl. **3**, 1–19 (2017)
4. Hardoon, D.R., Szedmak, S., Shawe-Taylor, J.: Canonical correlation analysis: an overview with application to learning methods. Neural Comput. **16**(12) (2014)
5. Cui, Y., Jiang, J., Lai, Z., et al.: Supervised discrete discriminant hashing for image retrieval. In: Pattern Recognition (2018)

6. Xu, X., Shen, F., Yang, Y., et al.: Learning discriminative binary codes for large-scale cross-modal retrieval. IEEE Trans. Image Process. **99**, 1 (2017)
7. Shen, F., Shen, C., Liu, W., et al.: Supervised discrete hashing. In: pp. 37–45 (2015)
8. Bouchard, G., Guo, S., Yin, D.: Convex collective matrix factorization. In: 144–152 (2013)
9. Rasiwasia, N., Costa, P.J., Coviello, E.: A new approach to cross-modal multimedia retrieval. In: Proceedings of the 18th ACM International Conference on Multimedia, pp. 251–260 (2010)
10. Zhen, Y., Yeung, D.-Y.: A probabilistic model for multimodal hashfunction learning. In: On knowledge Discovery and Data Mining. ACM, pp. 940–948 (2012)
11. Chua, T.-S., Tang, J., Hong, R. H.: Nus-wide: A real-world web image database from national university of singapore. In: International Conference on Image and Video Retrieval 48, vol. 9, pp. 1–48 (2009)
12. Zhou, J.: Latent semantic sparse hashing for cross-modal similarity search. In: Proceedings of the In ACM SIGIR Conference on Research and Development in Information Retrieval, pp. 415–424 (2014)
13. Ding, G., Guo, Y., Zhou, J.: Collective matrix factorization hashing for multimodal data. In: IEEE Conference on Computer Vision and Pattern Recognition, pp. 2083–2090 (2014)
14. Zhen, Y., Yeung, D.: Co-regularized hashing for multimodal data. In: Neural Information Processing Systems, pp. 1385–1393 (2012)
15. Zhang, D., Li, W.-J.: Large-scale supervised multimodal hashing with semantic correlation maximization. In: AAAI Conference on Artificial Intelligence (2014)

Sparse Discriminant Principle Component Analysis

Zhihui Lai, Mangqi Chen, Dongmei Mo, Xingxing Zou and Heng Kong

Abstract Sparse Principal Component Analysis (SPCA) is a regression-type optimization problem based on PCA. The main advantage of SPCA is that it can get modified PCs with sparse loadings so as to improve the performance of feature extraction. However, SPCA does not consider the label information of the data, which degrades its performance in some practical applications. To address this problem, we integrate the property of Least Squares Regression (LSR) into SPCA to use the prior knowledge to obtain the modified PCs with sparsity as well as discriminative information. Moreover, unlike LSR and its derivatives, the number of the modified PCs of SDPCA is not limited by the number of class, namely, SDPCA can address the small-class problem in LSR based methods. To solve the optimization problem, we also propose a new algorithm. Experimental results on product dataset, face dataset and character dataset demonstrate the effectiveness of SDPCA.

Keywords Pattern recognition · Feature extraction
Sparse principal component analysis · Least squares regression

1 Introduction

Principal Component Analysis (PCA) [1] and Linear Discriminant Analysis (LDA) [2] have a common drawback, that is, they take all the information of the data into consideration, which makes them lack the sparse representation of the data. Sparse

Z. Lai (✉) · M. Chen · D. Mo
College of Computer Science and Software Engineering,
Shenzhen University, Shenzhen 518060, China
e-mail: lai_zhi_hui@163.com

X. Zou
Institute of Textiles and Clothing,
The Hong Kong Polytechnic University, Hong Kong, China

H. Kong
School of Medicine, Shenzhen University, Shenzhen 518060, China

© Springer Nature Switzerland AG 2019
W. K. Wong (ed.), *Artificial Intelligence on Fashion and Textiles*,
Advances in Intelligent Systems and Computing 849,
https://doi.org/10.1007/978-3-319-99695-0_14

learning is considered to be important and effective in the real applications in pattern recognition and classification, because it can highlight the important features of the data and simultaneously ignore the useless information. The methods based on sparse learning can reduce the impact from noise or redundant information of the data. Lasso [3] and Elastic net [4] is the most representative methods in terms of sparse learning. Based on the elastic net, sparse PCA (SPCA) [5], sparse LDA (SLDA) [6] were proposed to extend PCA, LDA to sparse cases, respectively. Both of SPCA and SLDA use L1-norm on the regularization term to obtain sparsity for feature selection or extraction. Least Squares Regression (LSR) and Ridge Regression (RR) are widely used in data analysis and other cases [7]. Many methods based on LSR or RR are also developed to deal with different problems in feature selection [8].

Although all of the methods mentioned above are effective in feature selection or extraction in some degree, they still have some drawbacks. First, the methods based on PCA do not consider label information of the data. Second, the LSR based methods have the small-class problem, which means that the number of the projections is limited by the number of class. Due to the small-class problem, they cannot obtain enough projections for effective feature selection on the occasion when the dimension of the data is high but the number of class is small. Therefore, to release these problems, we need a new model to not only consider the label information of the data, but also release the small-class problem to obtain enough projection for effective feature selection. Base on this regard, in this paper, we propose a new method called Sparse Discriminant Principle Component Analysis (SDPCA). The contributions of this paper can be summarized as the following threefolds:

(1) The proposed SDPCA based on the well-known SPCA can use label information to obtain the principal component with sparsity and discriminative information to improve the performance of feature extraction.
(2) Unlike the methods based on LSR, the proposed SDPCA is able to break through the limitation of the small-class problem so as to obtain enough projections for more effective pattern recognition or classification.
(3) Experimental results on product dataset and face dataset demonstrate the good performance of the proposed SDPCA in terms of recognition tasks.

2   The Proposed Method

2.1   Sparse Discriminant Principle Component Analysis (SDPCA)

In this paper, all the matrices are denoted as bold uppercase italic letters, i.e., X, Y, etc., vectors are written as bold lowercase italic letters, i.e. x, y, etc., scalars are presented as lowercase italic letters, i.e., i, j, n.

Suppose the sample matrix is $X \in R^{d \times n}$, where d is the dimension of the data and n is the number of samples, namely, each column of X (i.e. x_i) represents a sample. The label matrix is denoted as $Y \in R^{n \times c}$, where c is the number of class. The elements Y_{ij} is set as 1 while x_i belongs to j-th class, otherwise, $Y_{ij} = 0$.

By integrating the property of SPCA and LSR to add lasso penalty to the regression criterion and consider the label information on the loss function, we have the following optimization problem of SDPCA

$$(A^*, P^*, C^*) = \arg \min_{A,P,C} \alpha \left\| Y - X^T P C \right\|_F^2 + (1 - \alpha) \sum_i^n \left\| x_i - A C^T P^T x_i \right\|_2^2$$

$$+ \beta \| P \|_F^2 + \sum_{j=1}^k \gamma_{1,j} \| p_j \|_1 \quad \text{s.t } A^T A = I_{c \times c}, C C^T = I_{k \times k} \quad (1)$$

where $X \in R^{d \times n}$ is the sample matrix, $Y \in R^{n \times c}$ is the label matrix, $P \in R^{d \times k}$ is the projection matrix containing k sparse PCs, $C \in R^{k \times c}$ and $A \in R^{d \times c}$ are auxiliary matrix as well as orthogonal matrix, α, β is the balance parameter, $\gamma_{1,j}$ is the parameter as in SPCA that penalizes the loadings of different principal components.

Similar to LSR, the first term in (1) incorporates label matrix to the loss function to use prior knowledge to obtain modified sparse PCs with discriminative information to improve the performance of feature extraction. The difference between SDPCA and LSR and its extensions is that SDPCA can obtain enough projections to improve the performance of feature extraction while the number of projection learned by LSR based methods is limited by the number of class. That is, the projection matrix of SDPCA is P with size of $d \times k$ and k is a variable that can be set as any integer so as to obtain enough projections for feature extraction. However, the size of projection matrix that learned from LSR based methods is $d \times c$, which means only c projections can be obtained for feature extraction. Therefore, SDPCA can break through the small-class problem that exists in LSR or its derivatives.

Compared with SPCA, SDPCA considers the label matrix on the loss function, which makes the projection matrix P contain discriminant information in some degree. Thus, the modified PCs of SDPCA are not only sparse but also discriminative.

2.2 The Optimization of SDPCA

There are three variables in (1) and they cannot be obtained directly since the objective function is not convex. Therefore, we need to optimize the optimization problem iteratively.

From (1), we have

$$\alpha \left\| Y - X^T PC \right\|_F^2 + (1 - \alpha) \sum_{i}^{n} \left\| x_i - A C^T P^T x_i \right\|_2^2 + \beta \left\| P \right\|_F^2 + \sum_{j=1}^{k} \gamma_{1,j} \left\| p_j \right\|_1$$

$$= \alpha tr(Y^T Y - 2 C^T P^T XY + C^T P^T X X^T PC) + (1 - \alpha) tr(X X^T$$

$$- 2 A^T X X^T PC + C^T P^T X X^T PC) + \lambda tr P^T P + \sum_{j=1}^{k} \gamma_{1,j} \left\| p_j \right\|_1 \qquad (2)$$

A step: Suppose the variables P and C are fixed (1) can be obtained by solving the following problem:

$$A^* = \arg \max_{A} (1 - \alpha) A^T X X^T P C A^T A = I \qquad (3)$$

Theorem 1 [5] *For a matrix $M \in R^{c \times k}$ with the rank of k and an orthogonal matrix $H \in R^{c \times k}$. The following optimization problem*

$$H = \arg \min tr(H^T M) s.t. H^T H = I_k \qquad (4)$$

can be solved by SVD of M, namely, $M = U D V^T$, then $H = U V^T$.

According to Theorem 1, we can know that the optimization problem of (3) can be solved by SVD of $(1 - \alpha) X X^T P C$, i.e.,

$$(1 - \alpha) X X^T P C = \tilde{U} \tilde{D} \tilde{V}^T \qquad (5)$$

Then, it goes

$$A = \tilde{U} \tilde{V}^T \qquad (6)$$

C step: When A and P are fixed, the optimal solution of (1) can be obtained by solving the following problem

$$C^* = \arg \max_{C} C^T P^T X \left(\alpha Y + (1 - \alpha) X^T A \right) C C^T = I_{k \times k} \qquad (7)$$

Similarly, (7) is optimized by

$$P^T X \left(\alpha Y + (1 - \alpha) X^T A \right) = \overline{U D V}^T \qquad (8)$$

then, we have

$$C = \overline{U V}^T \qquad (9)$$

P step: From (2), we have

$$P^* = \arg\min_{P} \alpha tr(Y^T Y - 2C^T P^T XY + C^T P^T XX^T PC) + (1-\alpha)tr(XX^T - 2A^T XX^T PC$$

$$+ C^T P^T XX^T PC) + \lambda tr(P^T P) + \sum_{j=1}^{c} \gamma_{1,j} \|p_j\|_1 \text{ s.t } A^T A = I_{c \times c}, CC^T = I_{k \times k} \quad (10)$$

It goes,

$$P^* = \arg\min_{P} tr(C^T P^T (XX^T + \lambda I)PC - 2C^T P^T X(\alpha Y$$

$$+ (1-\alpha)X^T A)) + \sum_{j=1}^{c} \gamma_{1,j} \|p_j\|_1 \text{ s.t } A^T A = I_{c \times c}, CC^T = I_{k \times k} \quad (11)$$

That is,

$$P^* = \arg\min_{P^*} tr(P^T (XX^T + \lambda I)P - 2P^T X(\alpha Y$$

$$+ (1-\alpha)X^T A)C^T) + \sum_{j=1}^{c} \gamma_{1,j} \|p_j\|_1 \text{ s.t } A^T A = I_{c \times c}, CC^T = I_{k \times k} \quad (12)$$

The optimization problem is equivalent to

$$P^* = \arg\min_{P^*} \|F - X^T P\|_F^2 + \beta \|P\|_F^2 + \sum_{j=1}^{c} \gamma_{1,j} \|p_j\|_1 \quad (13)$$

where $F = (\alpha Y + (1-\alpha)X^T A)C^T$.

Suppose $P^* = (p_1, p_2, ..., p_k)$ where p is the modified sparse PC. From (13), p can be obtained by solving the below problem

$$p^* = \arg\min_{p} \|F_i - X^T p\|_F^2 + \beta \|p\|_F^2 + \gamma \|p\|_1 \quad (14)$$

Since the optimization problem in (14) is similar to that of elastic net [9], the optimal p^* can be obtained by solving the elastic net problem.

3 Experiments

In this section, experiments on product dataset and face dataset are conducted to evaluate the proposed SDPCA. For comparison, the related PCA methods (i.e., PCA, SPCA [5], PCA-L1 [10]), the sparse discriminant method (i.e., SLDA [6]) and the LSR based method (i.e., DLSR [8]) are also tested on all the databases.

3.1 Datasets Description

The Fashion-MNIST dataset [11] contains a training set of 60,000 examples and a test set of 10,000 examples, demonstrating different aspects of the product [11]. 10% of the samples from this dataset are used in our experiments to evaluate the performance of SDPCA in dealing with more challenging classification tasks on real product.

The AR database [12] contains over 4000 face images from 126 people. 20 images from 120 individuals are selected and used in our experiments to evaluate the performance of SDPCA on the occasion when there are variations on the face with different facial expressions, lighting conditions, and occlusions.

3.2 Experimental Setting

In all experiments, PCA is first used as the preprocessing to reduce the dimension of the data. Then each method runs 10 times to perform feature extraction and the nearest neighbor (NN) classifier is used for classification. For SDPCA, the variables are randomly initialized because the performance is not sensitive to their initial values. The values of parameter α, β in SDPCA need to be selected from the area of $[-3,...,3]$ and $[10^{-3},\cdots,10^{3}]$ on all databases, respectively.

For the comparative methods, their parameters are set as the introduction in the original paper. For example, the parameter in SLDA, DLSR is set as $[10^{-3},...,10^{3}]$, $[10^{-4},...,10^{1}]$, respectively.

3.3 Experimental Results and Comparison

On Fashion-MNIST dataset [11], 6,000 images are used for training while 1,000 images are used for testing. On AR database, $l(l = 4, 5, 6)$ images of each class are randomly selected to form the training set while the rest are used for testing.

In the experiments, the optimal values of α and β on Fashion-MNIST dataset are $[1, 2, 3]$ and $[10^{-3},...,10^{3}]$ while the optimal values of α and β on AR dataset are $[-2, ..., 2]$ and $[10^{-3},...,10^{3}]$, respectively. The average recognition rates versus the dimension of different methods are presented in Fig. 1. Figure 2 shows the convergence curve of SDPCA while Tables 1 and 2 list the performance of all methods on Fashion-MNIST and AR dataset, respectively.

According to the experimental results, we have the following interesting points:

(1) On all experiments, the proposed SDPCA is able to obtain the best performance. The potential reason for this phenomenon is that SDPCA integrates the property of SPCA and LSR, with which it can not only obtain sparsity but also consider the label information for effective feature extraction.

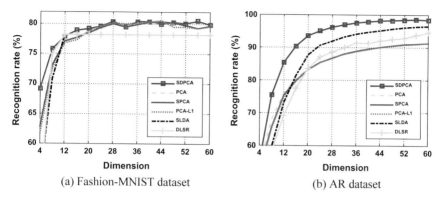

Fig. 1   The recognition rate versus the dimension

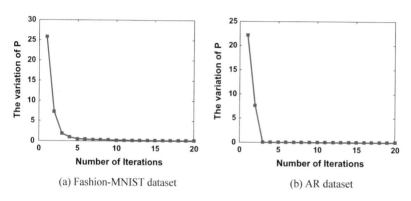

Fig. 2   The convergence curve of P

Table 1   On Fashion-MNIST dataset, the performance (recognition rate (%) and dimension) of all methods

Samples	PCA	SPCA	PCA-L1	SLDA	DLSR	SDPCA
6000	80.50 26	80.40 44	80.60 44	78.20 10	78.20 10	**81.10 34**

(2)   Compared with the PCA based methods (i.e. PCA, SPCA, PCA-L1), the advantage of SDPCA is that it considers label information on the loss function to obtain the PCs with not only sparsity but also discriminative information to improve the performance of feature extraction.

4   Conclusion

In this paper, we integrate the property of SPCA and LSR to propose a new method called Sparse Discriminant Principle Component Analysis (SDPCA) so as to not only

Table 2 On AR dataset, the performance (recognition rate, standard deviation (%), dimension) of all methods

l	PCA	SPCA	PCA-L1	SLDA	DLSR	SDPCA
4	91.30 ± 8.79 60	91.34 ± 8.97 60	91.30 ± 9.63 60	96.56 ± 11.43 58	94.55 ± 8.42 60	**98.58 ± 9.72 54**
5	92.94 ± 4.08 60	92.92 ± 4.09 60	92.94 ± 4.23 60	97.82 ± 6.89 60	95.83 ± 5.28 60	**99.15 ± 5.15 56**
6	93.17 ± 2.99 60	93.16 ± 3.07 60	93.17 ± 3.08 60	98.10 ± 5.00 60	96.58 ± 3.77 60	**99.18 ± 3.84 52**

obtain modified sparse principle components, but also consider the label information for feature extraction. Moreover, SDPCA can avoid the small-class problem that exists in LSR and its derivatives. Therefore, SDPCA can obtain enough projections to improve the performance of pattern recognition or classification. Experimental results on product dataset and face dataset show that SDPCA is superior to the related PCA methods (i.e., PCA, SPCA, and PCA-L1), the sparse discriminant method (i.e., SLDA) and the LSR based method (i.e., DLSR).

Acknowledgements This work was supported in part by the Natural Science Foundation of China (Grant 61573248, Grant 61773328, Grant 61773328 and Grant 61703283), Research Grant of The Hong Kong Polytechnic University (Project Code:G-UA2B), China Postdoctoral Science Foundation (Project 2016M590812 and Project 2017T100645), the Guangdong Natural Science Foundation (Project 2017A030313367 and Project 2017A030310067), and Shenzhen Municipal Science and Technology Innovation Council (No. JCYJ20170302153434048 and No. JCYJ20160429182058044).

References

1. Turk, M.: Eigenfaces for recognition. J. Cogn. Neurosci. **3**, 71–86 (1991)
2. Belhumeur, P.N., Hespanha, J.P., Kriegman, D.J.: Eigenfaces vs. fisherfaces: recognition using class specific linear projection. IEEE Trans. Pattern Anal. Mach. Intell. **19**, 711–720 (1997)
3. Tibshirani, R.: Regression shrinkage and selection via the Lasso. J. R. Stat. Soc. Ser. B Stat. Methodol. **58**, 267–288 (1996)
4. Zou, H., Hastie, T.: Regularization and variable selection via the elastic net. J. R. Stat. Soc. Ser. B. **67**, 301–320 (2005)
5. Zou, H., Hastie, T., Tibshirani, R.: Sparse principal component analysis. J. Comput. Graph. Stat. **15**, 1–30 (2004)
6. Qiao, Z., Zhou, L., Huang, J.Z.: Sparse linear discriminant analysis with applications to high dimensional low sample size data. IAENG Int. J. Appl. Math. **39**, 48–60 (2009)
7. Accelerator, S.L.: Regularized Discriminant Analysis. Publ. Am. Stat. Assoc. **84**, 165–175 (1989)
8. Xiang, S., Nie, F., Meng, G., Pan, C., Zhang, C.: Discriminative least squares regression for multiclass classification and feature selection. IEEE Trans. Neural Networks Learn. Syst. **23**, 1738–1754 (2012)
9. Zou, H., Hastie, T.: Regression shrinkage and selection via the elastic net, with applications to microarrays. J. R. Stat. Soc. Ser. B. **67**, 301–320 (2003)

10. Kwak, N.: Principal component analysis based on L1-norm maximization. IEEE Trans. Pattern Anal. Mach. Intell. **30**, 1672–1680 (2008)
11. Xiao, H., Rasul, K., Vollgraf, R.: Fashion-MNIST: a novel image dataset for benchmarking machine learning algorithms. 1–6 (2017)
12. Martinez, A.A., Benavente, R.: The AR face database. CVC Tech. Reptort #24. (1998)

New Product Design with Popular Fashion Style Discovery Using Machine Learning

Jiating Zhu, Yu Yang, Jiannong Cao and Esther Chak Fung Mei

Abstract Fashion companies have always been facing a critical issue to design products that fit consumers' needs. On one hand, fashion industries continually reinventing itself. On the other hand, shoppers' preference is changing from time to time. In this work, we make use of machine learning and computer vision technologies to automatically design new "must-have" fashion products with popular styles discovered from fashion product images and historical transaction data. Products in each discovered style share similar visual attributes and popularity. The visual-based fashion attributes are learned from fashion product images via a deep convolutional neural network (CNN). Fusing together with popularity attributes extracted from transaction data, a set of styles is discovered by Nonnegative matrix factorization(NMF). Eventually, new fashion products are generated from the discovered styles by Variational Autoencoder (VAE). The result shows that our method can successfully generate combinations of interpretable elements from different popular fashion products. We believe this work has the potential to be applied in the fashion industry to help to keep reasonable stocks of goods and capture most profits.

Keywords Fashion style discovery · Deep learning · VAE generator

J. Zhu · Y. Yang (✉) · J. Cao · E. C. F. Mei
The Hong Kong Polytechnic University, Hong Kong, China
e-mail: csyyang@comp.polyu.edu.hk

J. Zhu
e-mail: jtzhu@polyu.edu.hk

J. Cao
e-mail: csjcao@polyu.edu.hk

E. C. F. Mei
e-mail: esther88@me.com

© Springer Nature Switzerland AG 2019
W. K. Wong (ed.), *Artificial Intelligence on Fashion and Textiles*,
Advances in Intelligent Systems and Computing 849,
https://doi.org/10.1007/978-3-319-99695-0_15

121

1 Introduction

Fashion industries live on the cutting-edge of design, continually reinventing itself. Every season, fashion companies launch new styles into the market and try to be the trend-setters who produce most "must-have" products. Which product will become the "must-have" one? It is totally decided by the market. We define the "must-have" popular fashion product as the product that has high sales. However, customers' needs are difficult to interpret. Different people have different preferences and their taste may also change from time to time. How to design new styles that meet customers' needs is an open research question. Currently, fashion companies offer styles based on designers' choices, which rely on their knowledge of the social situation or cultural phenomena and guess what will be popular in the next season. However, the stakes are high, since unsold products at the end of a season are either sold at huge discounts or destroyed to protect the exclusivity of a brand name. In this paper, in order to complement the traditional investigation on future fashion trend, we aim to discover the style pattern of historical fashion products and consider customer's needs by machine learning and computer vision techniques. Variations or evolutions of design leading trend can be explored by learning through continuous selling items. We believe that information from image data goes beyond the knowledge boundary of human and captures more precise characteristics. With our method, new products could be automatically designed based on the discovered style pattern along with transaction information, such that they will have a higher probability to become popular and lead to more profits.

Specifically, given a set of fashion product images with transaction history, our objective is to discover the style of "must-have" fashion products and automatically generate new fashion products with "must-have" styles. The contribution of our work contains two-fold

- We propose a new approach to successfully discover popular fashion product styles from product images and transaction history without supervision.
- Our approach could automatically generate new "must-have" fashion products that may have higher probability to meet customers' needs than traditional designer choices.

Applying this work into fashion industries, fashion companies could launch new products into markets based on customers' needs, which would benefit targeted marketing incentives with higher successful rate, capture most profits and keep reasonable stocks of goods.

2 Related Works

In the fashion industry, the majority of sales forecasting models are using statistical methods [1], which however, depend on many manually designed attributes other than attributes extracted from images. Even with big data analytics, the fashion industry still extracts the feature from the product description and sales data [2] to infer the preferred styles and color for the next season.

Currently, there are several studies on customer preferences using image data, such as catalog image recommendation [3]. However, most of the recommendations are based on image retrieval, e.g., find similar clothes with a given image. This kind of recommendation neglects customers' preference in styles of products. To capture the market scale preference, modeling style characteristics has become a popular computer vision task (e.g., categorize images with similar latent features) nowadays [4]. However, the popularity of each fashion style is still missing. Therefore, to figure out fashion product styles along with their popularity, we extended the existing style discovery method [4] to incorporate both visual contents and popularity information. And to take a step forward, we generate new products based on the discovered styles, which are not attempted in previous fashion design studies using generative models [5].

3   A Machine Learning Approach for New Products Design

There are two challenges in solving the problem of automatically generating the next "must-have" fashion products. One challenge is how to discover fashion product styles sharing similar visual attributes and popularity. Popularity here refers to high sales. Another challenge is how to automatically generate new fashion products matching customers' preference. We proposed a new approach based on historical product images and transaction data. It first trains a CNN-based deep attribute model to learn the visual based fashion attributes from fashion product images and fuse together with popularity attributes extracted from transaction data. Then it discovers a set of styles that are sharing similar visual attributes and popularity from product images without supervision. Finally, new "must-have" fashion products are automatically designed by Variational Autoencoder (VAE) with images of discovered popular styles. The framework for the proposed approach is presented in Fig. 1.

To better extract the visual attributes from fashion product images, robust representations of fashion products need to be learned from images first and they should be interpretable in visual elements. Thus, a deep convolutional model would be trained for attribute prediction using the DeepFashion dataset [6], which contains more than

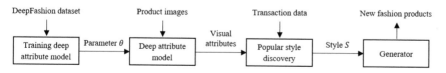

Fig. 1   The proposed framework for generating popular fashion products

200,000 images labeled with 1000 semantic attributes collected from online fashion websites.

Inspired by [4], we proposed a deep attribute model based on CNN. The model is composed of 11 convolutional layers with 3×3 filter size followed by three fully connected layers and two dropout layers with the probability of 0.5. Additionally, all the layers in our model are followed by a batch normalization layer and a rectified linear unit (ReLU) except the first two fully connected layers are followed by a scaled exponential linear unit (SeLU). We implemented our model in Keras with the weighted binary cross entropy loss to train the network for binary attribute prediction. The network is trained using SGD for stochastic optimization.

Given a set of images $= \{c_i\}_{i=1}^N$, the probability of M attributes for ith image is a_i. With the deep attribute model, we can get the trained model parameters θ and have $a_i = f_a(c_i|\theta)$. That is to say, the probability of semantic attributes such as floral print, stripe and arrow collar, etc., in a given fashion product image can be obtained, and we adopt it as the visual attributes for representing a given product image.

The deep attribute model yields a matrix $A \in \mathbb{R}^{\{M \times N\}}$ representing the probability of M visual attributes for N images. The sales data for each product are simply treated as an additional attribute. By augmenting matrix $A \in \mathbb{R}^{\{M \times N\}}$ to $\in \mathbb{R}^{\{(M+1) \times N\}}$, the sales information can also be considered for discovering latent styles. To discover the set of K latent styles $S = \{s_k\}_{k=1}^K$, we use a nonnegative matrix factorization (NMF) method to infer nonnegative matrices $W \in \mathbb{R}^{\{M \times K\}}$ and $H \in \mathbb{R}^{\{K \times N\}}$ such that $A \approx WH$. In this way, the discovered styles could share the similar visual attributes and popularity at the same time.

To produce products that are similar to those discovered styles but not the same, we formalize it as a problem that aims to learn a model $\hat{P}$ that is as similar as possible to P which is an unknown distribution where examples X follows [7]. Generative Adversarial Nets (GAN) [8] and Variational Autoencoder (VAE) [9] are two state-of-art machine learning generative models. GAN is to find the Nash Equilibrium between discriminator net from true distribution $P(X)$ and the generator net from model distribution $\hat{P}(X)$. VAE, on the other hand, is a probabilistic graphical model rooted in Bayesian inference. After comparing the performance between both, VAE is more suitable in our cases. It can generate combinations of interpretable elements from different products. This might because the scale of training samples is small and VAE can learn the latent variables that are interpretable.

4 Experiment Results

Our model is evaluated on a real-world data that is collected from a Hong Kong company. The data for the experiment contain 950 T-shirts images with 2058 transactions including online and offline purchases from the year 2015 to 2017.

Evaluation for Deep Attribute Model. We trained our deep attribute model on cropped upper body images from the DeepFashion dataset. Due to the nature of the T-shirts, we only consider upper body clothes. Therefore, we cropped upper body

Fig. 2 The product images for the discovered styles with high popularity

Fig. 3 The product images for the discovered styles with average popularity

images in the DeepFashion dataset and only use 533 attributes, which frequently appear in the upper body dataset (we take attributes that appear more than 100 times in this experiment). We split the DeepFashion dataset into 80% for training, 10% for validation and 10% for testing. Our attribute predictions average 77% AUC on a held-out DeepFashion test set.

Style Discovery. The NMF method in sklearn package in python has been used to learn $K = 30$ styles. For each style, we ranked top 10 images ordered by their attributes' similarities. As we can see in Fig. 2, the colors in style 1 tend to be cold and dark, while in style 10 the colors tend to be pinky. Two styles that have average popularity in Fig. 3 both have very complicated prints compared to the styles with higher popularity in Fig. 2. From this observation, we can say that customers in our data prefer simple and refreshing styles than the complicated design with detailed prints.

The average transaction for each product in our data set is 2.29. Hence, the transactions for a style (10 products) with average popularity should be 22.9. From Table 1, we can find that half of the styles we discovered have transactions higher than 22.9, which shows that our style discovery can find styles that have higher popularity.

"Must-have" Product Generation. The images from style 1 and style 10 are fed into the VAE with concrete latent distribution [10] respectively. As we can see from Fig. 4, the generated T-shirts have mixture elements from the original ones but are consistent with the original styles. Some obvious changes are: long sleeves to short sleeves, new color, the mixture of graphic prints and the design of the collar transfer from one shirt to another. For example, the generated shirt in the first column in style 1 from Fig. 4 contains the stripes and prints from the shirt in the fourth column in

Table 1 Total transactions for each discovered style

	1	2	3	4	5	6	7	8	9	10
Style	1	2	3	4	5	6	7	8	9	10
Sales	111	20	18	22	29	16	66	16	29	43
Popular	Yes	No	No	No	Yes	No	Yes	No	Yes	Yes
Style	11	12	13	14	15	16	17	18	19	20
Sales	16	17	33	15	25	11	77	14	54	15
Popular	No	No	Yes	No	Yes	No	Yes	No	Yes	No
Style	21	22	23	24	25	26	27	28	29	30
Sales	31	65	25	19	24	22	14	20	22	15
Popular	Yes	Yes	Yes	No	Yes	No	No	No	No	No

New products from style 1

New products from style 10

Fig. 4 Examples of new fashion products generated from style 1 and style 10

style 1 in Fig. 2, and the color mixture from shirts in the fifth and sixth column in style 1 in Fig. 2.

5 Conclusions

In this paper, we proposed a new approach for generating the next "must-have" fashion products. We first learn the visual attributes from an existing large dataset with fashion product images. Second, we incorporate visual attributes and popularity together from real-world product images with transaction data. We then discover a set of styles that are sharing similar visual attributes and popularity in an unsupervised manner. Finally, new "must-have" fashion products that have interpretable elements are automatically designed from images of discovered popular styles. With our method, by replacing transaction data to investigation data (votes on fashion products that will be in the market), we can also visualize designs that contain the fashion leading components. In the future, we are going to apply our method on a bigger dataset and evaluate the popularity of new generated products in the real world. The current dataset of generated products is not in a big variety. With more data, we would like to validate that our method can also generate more creative products from the discovered styles.

Acknowledgements The work is partially supported by the funding for Project of Strategic Importance provided by The Hong Kong Polytechnic University (Project Code: 1-ZE26), funding for the demonstration project on large data provided by The Hong Kong Polytechnic University (project account code: 9A5 V), and RGC General Research Fund, PolyU 152199/17E.

References

1. Brahmadeep, Thomassey, S.: Intelligent demand forecasting systems for fast fashion. In: Information Systems for the Fashion and Apparel Industry, pp. 145–161 (2016)
2. Banica, L., Hagiu, A.: Using big data analytics to improve decision-making in apparel supply chains. In: Information Systems for the Fashion and Apparel Industry, pp. 63–95 (2016)

3. Shankar, D., Narumanchi, S., Ananya, H.-A., Kompalli, P., Chaudhury, K.: Deep learning based large scale visual recommendation and search for e-commerce (2017). arXiv preprint arXiv: 1703.02344

4. Al-Halah, Z., Stiefelhagen, R., Grauman, K.: Fashion forward: Forecasting visual style in fashion. In: ICCV, pp. 388–397 (2017)

5. Deverall, J., Lee, J., Ayala, M.: Using generative adversarial networks to design shoes: the preliminary steps, CS231n in Stanford (2017)

6. Liu, Z., Luo, P., Qiu, S., Wang, X., Tang, X.: DeepFashion: powering robust clothes recognition and retrieval with rich annotations. In: Proceedings of IEEE Conference on Computer Vision and Pattern Recognition (CVPR), pp. 1096–1104 (2016)

7. Doersch, C.: Tutorial on variational autoencoders, (2016) arXiv preprint arXiv: 1606.05908

8. Goodfellow, I., Pouget-Abadie, J., Mirza, M., Xu, B., Warde-Farley, D., Ozair, S., Courville, A.-C., Bengio, Y.: Generative adversarial nets. In: Proceedings of NIPS, pp. 2672–2680 (2014)

9. Kingma, D.-P., Welling, M.: Auto-encoding variational Bayes. In: ICLR (2014)

10. Variational Auto Encoder with Concrete Latent Distribution. https://github.com/EmilienDupo nt/vae-concrete. Accessed 28 Mar 2018

Challenges in Knitted E-textiles

Amy Chen, Jeanne Tan, Xiaoming Tao, Philip Henry and Ziqian Bai

Abstract This paper considers the progress made in E-textiles within knitted textiles and discusses what "Project Jacquard" and the debut of the woven Levi's Commuter X Jacquard by Google jacket helps reveal about the relationship between E-textiles and textiles manufacturing. The paper considers research conducted within the fields of Art, Design and Technology, along with materials with interesting and novel properties that have been integrated into knitted textiles by practitioners and researchers. Such materials can embellish or enhance knitted fabric, from creating additional visual interest to practical functions. However, due to the physical properties of these types of materials, not all materials can be knitted into the fabric with ease; the optimal machine settings and techniques must be determined. Adapting to the physical characteristics of these innovative materials is a logical design requirement of the prototype development process but when we look to adopt the same principles as "Project Jacquard"; manufacturing knitted E-textiles to scale, the challenges of the material/machinery relationship become more of an issue. This raises the question as to whether it is better to develop the material for better textiles integration, or to optimize the production process to suit the material.

Keywords E-textiles · Photonic textiles · Polymeric optical fibers · Knitted fabric Electronics · Conductive yarn

1 Introduction

E-textiles has been steadily making its way into mainstream garments. Earlier E-textiles garments, such as the Galaxy dress by CuteCircuit [1] used hand embroidered circuity, while recently "The Marlene Project" [2] used the technique of e-broidery®

A. Chen (✉) · J. Tan · X. Tao · Z. Bai
Institute of Textiles and Clothing, Hong Kong Polytechnic University, Hong Kong, China
e-mail: amy.f.chen@connect.polyu.hk

P. Henry
School of Design, University of Leeds, Leeds, UK

© Springer Nature Switzerland AG 2019
W. K. Wong (ed.), *Artificial Intelligence on Fashion and Textiles*,
Advances in Intelligent Systems and Computing 849,
https://doi.org/10.1007/978-3-319-99695-0_16

to integrate lighting into the garment. Now, in a step towards everyday apparel, Google's Project Jacquard and Levi's have developed the Commuter X Jacquard jacket, which incorporates a woven "gesture-sensitive" fabric that provides the wearer with the opportunity to control their mobile phone through their jacket [3].

This practical functionality exemplifies E-textiles being incorporated into a fashionable garment but do not fully reflect the wider innovation in this area. Academic research shows a greater diversity in E-textiles, with the development of sensors, actuators and power sources. A number of different textiles manufacturing techniques have been utilized to create E-textiles, such as handcraft techniques like crochet and felting [4], fabric printing [5] and dyeing [6], knitting [7], as well as weaving [8] and embroidery.

While advanced knitting technologies are now being used by fashion and sports brands, i.e., seamlessly knitted garments and footwear by Uniqlo and Nike respectively, knitted E-textiles is not present in more mainstream garments, despite the progressive academic research in this area. Current literature presents the challenges and solution for knitted E-textiles at a small scale, consequently, there is a limit to the applicability of those design principles to larger scale production. Therefore, it is useful to analyze Google's "Project Jacquard", as it provides an insightful case study for scalable E-textiles integration.

2   Knitted E-textiles

Innovative materials can be more easily integrated into fabric when their properties are very similar to that of conventional yarns. Conductive yarns demonstrate this principle. In contrast, Polymeric Optical Fibers (POF) have very different physical properties to conventional yarn. They are more rigid and structurally different to yarns. While it might be assumed that conductive yarn can be knitted into a garment with ease, there are still issues that need to be considered during the design process.

2.1   Polymeric Optical Fibers

Polymeric optical fibers (POF) can be used for sensing and illumination purposes. Fiber Bragg grating (FBG) is one of the techniques that is used to create POF sensors, and existing reviews on the subject have discussed its use as a strain, temperature or humidity sensor [9]. FBG works in that the technique changes the refractive index in the fiber's core. Optical signals are transmitted through the fiber and reflected by the FBG. When there is a change in conditions, i.e. temperature change, this affects the wavelength of the light that is reflected. The POF acts as a sensor by measuring the change in the reflected light.

POF are not suited to being knitted into a fabric, as the tight bends of the knitted structure can cause excessive damage, preventing light from traveling along the full length of the fiber. On the other hand, a certain amount of damage to the fiber is necessary for illumination via the lateral side of the fiber, and further processes, like laser engraving, can be used to determine the light emission levels [10]. To overcome the challenges of knitting with POF, Inlay has been used to incorporate POF into knitted fabric. The POF is held in place by the knitted loops, rather than being knitted into the fabric and this is accomplished through either hand or machine manipulation. By using Inlay, a range of fiber diameters can be used, with the cited examples using POF from 0.25 to 0.75 mm.

Inlaying can place some restrictions on the design of the fabric and garment. The knit POF samples produced as part of CraftTech [11] show that patterns can be created by manipulating the knit structure to reveal select areas of POF. Nevertheless, the POF is still integrated horizontally, which impacts the drape of the fabric, and subsequently the silhouette of the garment [12]. Yet, the rigidity of POF can be used to its advantage, as seen in the sculptural knitted art pieces by Blomsedt [13]. While POF requires the additional step of attaching the POF to the light source, one benefit of this material is that a large section of fabric can be illuminated by using one light emitting diode.

2.2 Conductive Yarn

Compared to POF textiles, there are a greater number of examples of E-textiles using conductive yarns. This is potentially due to conductive yarns being easier to integrate, and the wide variety of conductive fibers and yarn available. Conductive yarns utilize the knitted structure to perform sensing and actuating functions. Stretch sensors functionality uses the change in resistance caused by stretch of the fabric, while the stroke sensing fabric detects the action by behaving in the same way as a switch in an electrical circuit [4]. Heated fabric can also be produced using conductive yarn [14], which dissipates the electrical power through the conductive yarn as heat. Spacer fabric has been adopted in the development of energy harvesting textiles. The energy harvesting fabric generates electricity as a result of pressure applied upon the fabric, which consists of electro-spun piezoelectric yarn knitted as spacer yarn, with conductive and non-conductive yarn used for the fabric faces [15].

Conductive yarn can be knitted into a range of fabric structures, given that they are the appropriate weight for the machine gauge. However, the development in conductive yarns focuses on its technical applications as opposed to the look and feel. Conductive yarns are predominantly gray or silver, and this can impact on the garment design decisions. For examples, in the heated garments produced by Chu Yin Ting, "the conductive paths were designed to match with the parallel vertical stripes … so that the paths become part of the pattern design" [16]. For spacer fabric, which has a distinctive draping quality, it is necessary to consider the impact of its placement on the garment and the effect on the garment form. While conductive yarns have overcome the technical difficulties, it presents some challenges when we look at integrating it into the design of a garment.

3 Google Jacquard: E-textiles at Scale

A key aim of the Project Jacquard was to develop solutions to "enable employing invisible ubiquitous interactivity at scale" [17]. Scalability is important as E-textiles moves beyond the realms of occasion wear and into everyday wear. Nevertheless, it is difficult to pursue economies of scale without the demand for the product [18].

3.1 Textiles Industry Requirements

During the development process of Project Jacquard, the same observation was made regarding the aesthetic limitations of conductive yarn. The solution that was developed was a conductive yarn consisting of conventional yarn braided around copper core, designed to have a "natural look and feel" [17]. The braided element of the yarn allows for a greater variety of colors and textures. The yarn has been successful used in industrial looms. Based on this, it suggests that the Jacquard yarn could work within a knitted structure, given that it has similar physical properties to conventional yarn.

Project Jacquard developed gesture-sensitive fabric and assembled into a garment through conventional cut and sew garment construction in collaboration with fashion designers. Cut and Sew is a knitwear construction technique is comparable woven garment construction, but there are other garment construction methods. Fully Fashioned consists of knitting garment pieces to shape before assembly, and in Whole Garment/Seamless knitting, the whole garment is knitted on the machine [19]. There are advantages and disadvantages to each method. Whole Garment knitting produce less waste, saving on material cost, and the seamlessness of the garment reduces labor cost while being more comfortable. However, this newer technology comes at a higher cost. In contrast, Cut and Sew is the most common production method, and does not require computerized knitting machines. As the fabric is cut into shape, there is waste, and further labor costs for garment assembly. The estimates on cost are further complicated by the fact that the cost varies depending on the style of the garment [20]. The relationship between knitted E-textiles and construction technique requires some investigation to strike a balance between cost-effectiveness and the envisioned end-product.

3.2 Connecting Electronics to Textiles

Electronics and textiles have different connection methods, with soldering for electronics and sewing for fabrics. Processes currently being used in commercial E-textiles include computerized embroidery and pre-made textiles cables [21]. Hand processes have been used in small scale work [22]. Project Jacquard took that approach initially but it was considered "laborious and error-prone" [17]. Optimization is necessary for scalable manufacturing. For Project Jacquard, the connection process was resolved through 3D weaving. The fabric is woven "as a two-layer textile, where the conductive (red) yarn forms a localized square conductive area in the top layer, then passes through the fabric to the bottom layer, where it floats. There, the yarns are free from the textiles and can be addressed individually" [17]. The Jacquard yarns are then connected using traditional soldering processes. Creating floats is possible in knitted fabric, as seen in Fairisle knitted fabric. However, woven fabrics can produce floats horizontally and vertically, while knitted fabric only produces horizontal floats, which may be less versatile. Another design element that could be taken forward is the use of snaps for removable electronics, as seen in the Google and Levi's jacket, the garments by Chu Yin Ting, and the textiles cables by Interactivewear [21].

A potential issue for knitted fabric is its extensibility. The weight of the electronics could cause unwanted stretch. This is not a problem for the Google and Levi's Jacket, which uses denim but could pose a problem for a ribbed knit fabric. Textiles cabling may not be suited to some knitted fabrics if they do not possess the same level of extensibility, or if the garment is thin, making the cabling visible.

4 Further Considerations

The lessons learned from Project Jacquard can be transferred into knitted E-textile using conductive yarns, as many of the issues raised within Project Jacquard can be found in knitted conductive fabrics. Due the differences in both material and technique, the same cannot be said for POF. Jacquard yarn was designed to fit into the industrial weaving process and used existing weaving techniques to overcome connectivity issues. For POF, the manufacture process has been adjusted to suit the material's challenging physical properties. Even though POF is difficult to knit, it has been successfully integrated into knitted fabric using an Inlay capable computerized knitting machine, showing that there is potential for scalable production. On the other hand, the attachment processes for conductive yarn is also not directly transferable to POF, as POF carries light rather than electricity, and it cannot be heated to a high temperature [23]. At present, the fibers are bundled and attached to the light emitting diode by hand [10], reminiscent of Project Jacquard's hand process. Key questions for further POF textiles development are whether it is better to develop the material for better textiles integration, or to optimize the production process to suit the material. By redeveloping the material to improve its compatibility with existing knitting technologies, it may be more easily adopted as manufacturers, in their existing production setup, rather than them having to invest in new knitting equipment. With that being said, manufacturing E-textiles garments is not solely the matter of producing E-textiles fabric. As investment is required to process other elements of the garment, such as the electronic connections, it can be argued that investment in new technology is inevitable. With the knitting industry being so diverse in it production technologies and techniques, improving the scalability of knitted E-textiles may require a multifaceted approach.

References

1. Cutecircuit (2009) The Galaxy Dress. http://cutecircuit.com/collections/the-galaxy-dress/. Accessed 27 Nov 2016
2. Elektrocouture.: The Marlene Project - ElektroCouture | Bespoke Elektronic Fashion Technologies. Elektrocouture. (2017) https://elektrocouture.com/the-marlene-project/. Accessed 14 Sept 2017
3. Levi's: Levi's® Commuter™ Trucker Jacket with Jacquard™ by Google. Levi's®. (2018). http://www.levi.com/US/en_US/womens-clothing-jackets-vests/p/287720000
4. Perner-Wilson, H., Buechley L., Satomi, M.: Handcrafting textile interfaces from a kit-of-no-parts. In: Paper presented at the Proceedings of the fifth International Conference on Tangible, Embedded, and Embodied Interaction, Funchal, Portugal (2011)
5. Karim, N., Afroj, S., Malandraki, A., Butterworth, S., Beach, C., Rigout, M., Novoselov, K.S., Casson, A.J., Yeates, S.G.: All inkjet-printed graphene-based conductive patterns for wearable e-textile applications. J. Mater. Chem. C **5**(44), 11640–11648 (2017). https://doi.org/10.1039/c7tc03669h
6. Fugetsu, B., Sano, E., Yu, H., Mori, K., Tanaka, T.: Graphene oxide as dyestuffs for the creation of electrically conductive fabrics. Carbon **48**(12), 3340–3345 (2010). https://doi.org/10.1016/j.carbon.2010.05.016
7. Wang, J., Long, H., Soltanian, S., Servati, P., Ko, F.: Electromechanical properties of knitted wearable sensors: part 1—theory. Text. Res. J. **84**(1), 3–15 (2014). https://doi.org/10.1177/0040517513487789
8. SubTela, S.: White wall hanging. (2007). http://subtela.hexagram.ca/Pages/White%20Wall%20Hanging.html
9. Zeng, W.: Polymer optical fiber for smart textiles. In: Tao X. (ed.) Handbook of Smart Textiles. Springer Singapore, Singapore, pp 109–125. (2015). https://doi.org/10.1007/978-981-4451-45-1_23
10. Tan, J.: Photonic Fabrics for Fashion and Interior. In: Tao X. (ed.) Handbook of Smart Textiles. Springer Singapore, Singapore, pp 1005–1033. (2015). https://doi.org/10.1007/978-981-4451-45-1_29
11. Tan, J., Toomey, A.: CraftTech: Hybrid Frameworks for Smart Photonic Materials. Royal College of Art, UK (2018)
12. Chen, A.: Literature Review and Research Methodology. Creating an Effective E-textiles Toolkit for Fashion Design. Manchester Metropolitan Univerisity, Manchester (2017)
13. Blomstedt, B.: LUX: Exploring Interactive Knitted Textiles Through Light and Touch. University of Boras (2017)
14. Mbise, E., Dias, T., Hurley, W.: 6 - Design and manufacture of heated textiles. In: Electronic Textiles. Woodhead Publishing, Oxford, pp 117–132. (2015). https://doi.org/10.1016/B978-0-08-100201-8.00007-2
15. Soin, N., Shah, T.H., Anand, S.C., Geng, J., Pornwannachai, W., Mandal, P., Reid, D., Sharma, S., Hadimani, R.L., Bayramol, D.V., Siores, E.: Novel 3-D spacer all fibre piezoelectric textiles for energy harvesting applications. Energy Environ. Sci. **7**(5), 1670–1679 (2014). https://doi.org/10.1039/c3ee43987a
16. Chui, Y.T.: Creation of wearable electronic clothing addressing end-user needs. Hong Kong: Institute of Textiles and Clothing, The Hong Kong Polytechnic University, Hong Kong (2017)
17. Poupyrev, I., Gong, N.-W., Fukuhara, S., Karagozler, M.E., Schwesig, C., Robinson, K.E.: Project jacquard: interactive digital textiles at scale. In: Paper presented at the Proceedings

of the 2016 CHI Conference on Human Factors in Computing Systems, San Jose, California, USA (2016)
18. Röpert, A.: Smart Textiles—How to Enter the Market. Berlin (2017)
19. Power, J.: 9—Developments in apparel knitting technology A2—Fairhurst, Catherine. In: Advances in Apparel Production. Woodhead Publishing, pp. 178–196. (2008). https://doi.o rg/10.1533/9781845694463.2.178
20. StitchWorld.: Flat Knit Production A Comparative Analysis. Stitch World (2010)
21. Mecnika, V., Scheulen, K., Anderson, C.F., Hörr, M., Breckenfelder, C.: 7—Joining technologies for electronic textiles A2—Dias, Tilak. In: Electronic Textiles. Woodhead Publishing, Oxford, pp. 133–153. (2015). https://doi.org/10.1016/B978-0-08-100201-8.00008-4
22. Peppler, K., Sharpe, L., Glosson, D.: E-textiles and the new fundamentals of fine art. In: Buechley, L., Peppler, K., Eisenberg, M., Kafai, Y. (eds.) Textile Messages: Dispatches from the World of E-textiles and Education. Peter Lang Publishing Inc, New York (2013)
23. Bai, Z.: Innovative photonic textiles: the design, investigation and development of polymeric photonic fiber integrated textiles for interior furnishings. Institute of Textiles and Clothing, The Hong Kong Polytechnic University, Hong Kong (2015)

The CF+TF-IDF TV-Program Recommendation

Li Yan, Cui Jinrong, Xin Liu, Yu JiaHao and He Mingkai

Abstract This paper first analyzes and discusses the traditional methods used in the TV-program recommendation system. Current studies show that the explicit data methods (user interest preference matrix) used in the real-world datasets does not work well and the precision of implicit data method (collaborative filtering) greatly relies on the data amount of each user. Then, we introduce a new method called CF+TF-IDF, which combines the collaborative filtering and TF-IDF algorithm. In order to analyze users' preference, we also add k-means++ algorithm in it to cluster the users. The core of the method is to infer users' preference from their viewing habits and the program type they choose. By using CF+TF-IDF, we build a TV-program recommendation model, aiming at improving users' viewing experience.

Keywords TV-program recommendation · Collaborative filtering · TF-IDF
k-means++

1 Introduction

Many works have been done to improve the TV-program recommendation system. In the past, proposals are mainly focused on personal recommendations based on the profile of a particular consumer. For example, MovieLens system [1] is based on an individual's taste as inferred from ratings and collaborative filtering. However, Masthoff [2] suggested that television viewing is often a group experience because people watch TV with their families or friends. And the TV-program recommendation systems are usually more useful for groups of people with similar interests, maybe the case of friends (homogeneous groups) rather than the case of families

L. Yan · X. Liu · Y. JiaHao · H. Mingkai
South China Agricultural University, Guangzhou 510642, China

C. Jinrong (✉)
College of Mathematics and Informatics, South China Agricultural University,
Guangzhou 510642, China
e-mail: tweety1028@163.com

© Springer Nature Switzerland AG 2019
W. K. Wong (ed.), *Artificial Intelligence on Fashion and Textiles*,
Advances in Intelligent Systems and Computing 849,
https://doi.org/10.1007/978-3-319-99695-0_17

137

(heterogeneous groups). Therefore, more research have been carried out to improve or create the group systems. For example, Chorianopoulos [3] proposed the application of recommendation methods within small networks of affiliated groups of TV viewers. However, no matter the system is mainly focused on personal recommendations or group, they all need to infer users' preference either from the explicit data input by users or the implicit data of users' viewing habits. Therefore, in this article, we will analyze and discuss traditional methods and thus create a new method calling CF-TF+IDF.

2   Previous Work

2.1   The Explicit Data Method

Based on the theory proposed by He [4] in 2016, we can build the user interest preference matrix from the users' viewing habits, including their viewing behaviors and time on requested TV-program. And according to Luo, the method can effectively profile the user preference and improve the precision of the recommendation. Therefore, here, we can repeat his research to see the performance of this method.

As we can see from Table 1, the range and the CV (Coefficient of Variation) of user final preferences is so big that leads to a conclusion that the explicit data method proposed by LuoHe cannot work well in the real-world datasets.

2.2   The Implicit Data Method

A common task of recommendation systems is to improve customer's experience through personalized recommendations based on prior implicit feedback.

Collaborative Filtering (CF) relies only on user's past behavior without requiring the creation of explicit profiles [5]. Recommendation systems rely on different types of input. This recommenders can infer user preferences from abundant implicit feedback, which indirectly reflects opinion through observing user behavior.

We reserve special indexing letters for distinguishing users from items: for users u, v, and for TV programs i, j. The input observable data associate users and TV programs through r_{ui} values. Here, we set the rule for r_{ui}:

Table 1 Descriptive statistics of user final preferences

Descriptive statistics of user final preferences							
Count	Mean	Std	Min	Max	Range	CV	IQR
3931	27.7	416.37	4.20E-08	21559.61831	21559.618	15.04453	1.251168

$$r_{ui} = \begin{cases} 1 & \text{both have request and purchase records} \\ 1 & \text{only have request records} \\ 0 & \text{only have purchase records} \\ 0 & \text{either have request nor purchase records} \end{cases} \tag{1}$$

In order to formalize the notion of confidence which the r_{ui} variables measure, we introduce a set of binary variables p_{ui} which indicates the preference of user u to TV program i. The p_{ui} values are derived by binarizing the r_{ui} values:

$$p_{ui} \begin{cases} 1 & r_{ui} > 0 \\ 0 & r_{ui} = 0 \end{cases} \tag{2}$$

We introduce a set of variables, C_{ui}, which measures our confidence for observing p_{ui}.

$$C_{ui} = 1 + r_{ui} \tag{3}$$

Here, we found that the result is better when $\alpha = 38$.
Where,

$$\begin{cases} x_u \in R^f, \text{ vector of user } u \\ y_i \in R^f, \text{ vector of program } i \\ p_{ui} = x_u^T y_i, \text{ the preferences of user } u \text{ towards program } i \end{cases} \tag{4}$$

Therefore, factors are computed by minimizing the following cost function:

$$\min_{x_*, y_*} \sum_{u,i} C_{ui}(p_{ui} x_u^T y_i)^2 + \lambda \left(\sum_u \|x_u\|^2 + \sum_i \|y_i\|^2 \right) \tag{5}$$

Minimizing the cost function for our user u, we have

$$x_u = (Y^T C^u Y + \lambda I)^{-1} Y^T C^u p(u), \tag{6}$$

which is equal to

$$x_u = (Y^T Y + Y^T (C^u - 1)Y + \lambda I)^{-1} Y^T C^u p(u), \tag{7}$$

We can derive a similar equation for program i:

$$y_i = (X^T X + X^T (C^i - 1)X + \lambda I)^{-1} X^T C^i p(i) \tag{8}$$

The whole process scales linearly with the size of the data. After computing the user-and item-factors, we recommend user u and the K available programs based on the largest value of

$$\widehat{p_{ui}} = x_u^T y_i, \tag{9}$$

where $\widehat{p_{ui}}$ symbolizes the predicted preference of user u for program i.

Here, we set $K = 6$.

$$\text{precision} = \frac{\text{the number of program both in recommended and actual result}}{\text{the number of recommend}} \tag{10}$$

From Fig. 1, we can see that the precision of the test set varies greatly, which is due to the inconsistent amount of each user's viewing habits data. Due to the difference in the number of viewing habits records of each user, the precision of the test sets are greatly different and the average of total test set precision is only 12%.

The more the amount of users' viewing history data, the better TV-program recommendation system works. But, a user cannot find out whether the system work well or not before he/she put effort in it, which called cold start problem [6]. Therefore, Baudisch [7] proposed a retrieval-oriented system called TV Scout which can support users in planning their personal TV consumption. Although, such system can avoid "cold-start" problem of information filtering systems by giving an immediate result to users, the operation of them are more difficult than the traditional one. Therefore, in this paper, we aim to reduce the "cold-start" problem instead of avoiding it.

```
Iterations 0... Training Loss 24969.37... Train Precision 0.013... Val Precision 0.000
Iterations 10... Training Loss 4464.67... Train Precision 0.520... Val Precision 0.067
Iterations 20... Training Loss 3362.28... Train Precision 0.653... Val Precision 0.053
Iterations 30... Training Loss 2826.39... Train Precision 0.760... Val Precision 0.027
Iterations 40... Training Loss 2462.81... Train Precision 0.773... Val Precision 0.027
Iterations 50... Training Loss 2158.85... Train Precision 0.760... Val Precision 0.027
Iterations 60... Training Loss 1846.44... Train Precision 0.747... Val Precision 0.027
Iterations 70... Training Loss 1410.68... Train Precision 0.747... Val Precision 0.027
Iterations 80... Training Loss -460.11... Train Precision 0.667... Val Precision 0.040
Iterations 90... Training Loss -78665.43... Train Precision 0.347... Val Precision 0.040
Iterations 100... Training Loss -1089850.75... Train Precision 0.093... Val Precision 0.053
Iterations 110... Training Loss -5076765.00... Train Precision 0.053... Val Precision 0.027
Iterations 120... Training Loss -14369247.00... Train Precision 0.053... Val Precision 0.027
Iterations 130... Training Loss -31138206.00... Train Precision 0.053... Val Precision 0.027
Iterations 140... Training Loss -57353572.00... Train Precision 0.053... Val Precision 0.027


Test Precision0.120
```

Fig. 1 Precision of the test set

3 CF+TF-IDF Model

In real applications of collaborative filtering, the rating matrix is usually very sparse, causing CF-based methods to degrade significantly in recommendation performance [8–11].

3.1 TF-IDF Model

First, we use TF-IDF (term frequency–inverse document frequency) to analyze the name of TV program and their plot summary in order to build the TF-IDF text similarity model which is based on the Word Segmentation. The model function is query indexes. Namely, when we input the plot summary, we can make a similar query of it and find out the other similar TV program. The result is sorted by the similarity from high to low.

Formula for non-normalized weight of term i in document j in a corpus of D documents is given as follows:

$$\text{weight}_{i,j} = \text{frequency}_{i,j} \times \log_2 \frac{D}{\text{document\_freq}_i}, \tag{11}$$

$$\text{idf} = \text{add} + \log_{\text{log\_base}} \frac{\text{totaldocs}}{\text{doc\_freq}} \tag{12}$$

The variable: doc_freq means document frequency and totaldocs stands for total number of documents. Here, log_base = 2, add = 0.

In this article, the number of words in dictionary is 21,055, the number of documents is 1794, and the number of words is 71,307. Therefore, it is difficult for us to judge the importance and representativeness of words and quantify their weight.

3.2 Our Work

In this article, we introduce the new approach is to combine the CF recommendation model with the TF-IDF text similarity model.

CF+TF-IDF model:

1. Collaborative Filtering the user historical viewing habits data in order to obtain the indirectly reflect opinion of each user. Namely, Collaborative Filtering helps us to judge the importance, characteristic and representative words in the program name and plot summary.
2. Obtaining the TF-IDF text similarity model through the word segmentation of program name and plot summary.

3. Through the Collaborative Filtering, we obtain top 3 recommended results of each user. First, segmenting words in each program's name and plot summary. Then, make a similar query of them through TF-IDF text similarity model. In this article, we set the minimum of similarity be 5%. The result is sorted descendingly. Here, we build the CF+TF-IDF recommendation program model. And, the results are called the recommended program options set.
4. The k-means++ and TF-IDF clustering texts are used to aggregate users' viewing history of TV-program label for building the similar preference user grouping model. The results are called user group sets.
5. In order to improve the Collaborative Filtering precision of users with less data, in this article, the user with the least record in each group was selected to conduct Collaborative Filtering in the group. Then repeat step 3.

3.3   Results

For the users with enough viewing history, we can directly recommend TV-program for them. Here, we take an example for the user with device number 10036, the top three recommended TV-program obtained by Collaborative Filtering is: "Oh My Grad", "Agent Princess", "Go Fighting", "The Tofu War", "The First Half of My Life".

Then, we input the top three recommended results into the CF+TF-IDF recommendation program model. The recommended program options set is shown below:

As we can see from Table 2, there are 23 TV-programs' text similarity. This quantity is 5% higher than that of the recommended program options set. At the same time, through the CF+TF-IDF recommendation program model, we obtain the priority of the recommended results. Hence, CF+TF-IDF recommendation program model is better than only using the CF (Collaborative Filtering) model.

Then we use k-means++ proposed by David Arthur and TF-IDF clustering texts to aggregate all classified categories which have been watched in order to build the similar preference user grouping model.

However, we use TF-IDF to cluster texts instead of making similar query. Therefore, we need to adjust the formulas of TF-IDF.

Table 2 Recommended program options set

Similarity	Program name	Priority
1.0000005	Agent Princess	1
1	The First Half of My Life	2
0.5420288	Go Fighting	3
…	…	…
0.05567534	Lost Love in Times	23

Table 3 User group sets

User group sets						
Favorite TV-program label	Drama	Cartoon	Action	Comedy	...	Disaster
Number of people (group)	246	56	53	52	...	1

Table 4 Recommended program options set (less data user)

Device	Group	Program name
10931	0	To the Sky Kingdom, Day and Night, Qinlang
11253	0	Give Me Five, Trump Card season, Taihang Hero
...	...	...
10186	0	The House That Never Dies, Avatar, The Unholy Alliance, Line Walker

$$\mathrm{idf}(t) = \log \frac{1 + n_d}{1 + \mathrm{d}f(d,t)} + 1 \tag{13}$$

In Eq. (13), n_d is the total number of documents, and $\mathrm{d}f(d,t)$ is the number of documents that contain term t. The resulting tf-idf vectors are then normalized by the Euclidean norm. The results of similar preference user grouping model are as shown in Table 3.

$$v_{\mathrm{norm}} = \frac{v}{||v||^2} = \frac{v}{\sqrt{v_1^2 + v_2^2 + \cdots + v_n^2}} \tag{14}$$

Here, we take the users of Group 0 as an example, the recommended program options set is shown below (Table 4):

4 Conclusion

In this paper, first, we find the implicit data method (Collaborative Filtering) has a delayed benefit problem. Therefore, we introduce a new method called the CF+TF-IDF. This method uses Collaborative Filtering to analyze the implicit data (users' viewing habits) and TF-IDF algorithm to analyze the explicit data (users' favorite TV-program label). By gathering the same view interest of the user in the same group, we obtain the homogeneous groups. Hence, we reduce the "cold start" problem by referring the viewing interest of other users in the same group. In this paper, we use the real-world datasets to test the feasibility and effectiveness of our work. The test results show that the proposed method improves the users' viewing experience in TV-program.

References

1. Yu, Z., Zhou, X., Hao, Y., Gu, J.: TV program recommendation for multiple viewers based on user profile merging. User Model. User-Adap. Inter. **16**, 63–82 (2006)
2. Masthoff, J.: Group modeling: selecting a sequence of television items to suit a group of viewers. User Model User-Adap. Inter. **14**, 37–85 (2004)
3. Chorianopoulos, K.: Personalized and mobile digital TV applications. Multimedia Tools Appl. **36**(1–2), 1–10 (2008)
4. He, L., Zhao, P.: An approach of building user interest perference model for TV viewers. Chin. J. Manag. Sci. **11**(24), 43–48 (2016)
5. Goldberg, D., Nichols, D., Oki, B., M., Terry, D.: Using collaborative filtering to weave an information tapestry. Commun. ACM **35**, 61–70 (1992)
6. Resnick, P.H.: Varian (Eds.). Special issue on recommender systems. Commun. ACM **40**(3), 56–89 (1997)
7. Baudisch, P., Brueckner, L.: TV Scout: Lowering the entry barrier to personalized TV program recommendation. Int. Conf. Adapt. Hypermedia Adapt. Web-based Syst. **3379**, 58–68 (2002)
8. Dong, X., Yu. L., Wu, Z., Sun, Y., Yuan, L., Zhang, F.: A hybrid collaborative filtering model with deep structure for recommender systems. In: Proceedings of the Thirty-First AAAI Conference on Artificial Intelligence, pp. 1309–1315 (2017)
9. Jia, W., Li, S., Li, X., Liu, B.: Collaborative filtering recommendation algorithm based on discrete quantity and user interests. Approach Degree **1**(44), 226–237 (2018)
10. Najafabadi, M.K., Mahrin, M.N., Chuprat, S., Sarkan, H.M.: Improving the accuracy of collaborative filtering recommendations using clustering and association rules mining on implicit data. Comput. Hum. Behav. **67**, 113–128 (2017)
11. Belacel, N., Durand, G., Leger, S., Bouchard, C.: Splitting-merging clustering algorithm for collaborative filtering recommendation system. In: International Conference on Agents & Artificial Intelligence, pp. 165–174 (2018)

Minimize the Cost Function in Multiple Objective Optimization by Using NSGA-II

Hayder H. Safi, Tareq Abed Mohammed and Zena Fawzi Al-Qubbanchi

Abstract This study proposes a new framework to minimize the cost function of multi-objective optimization problems by using NSGA-II in economic environments. For multi-objective improvements, the most generally used developmental algorithms such as NSGA-II, SPEA2 and PESA-II can be utilized. The economical optimization framework includes destinations, requirements, and parameters which continuously can change with time. The minimization of the cost function issue is one of the most important issues as in the case of stationary optimization problems. In this paper, we propose a framework that can possibly reduce the high cost of all functions that used in economic environments. Our algorithm uses a set of linear equations as inputs which depend on multi-objective algorithm that based on a Non-Dominated Sorting Genetic Algorithm (NSGA-II). The results of our experimental study show that the proposed framework can efficiently be used to reduce the cost and time of optimizing the economical problems.

Keywords Cost function · NSGA-II multi-objective problem
Vector optimization · Vehicle suspension system

H. H. Safi (✉) · T. A. Mohammed
Altinbas University Istanbul, Istanbul, Turkey
e-mail: haydersafi6@gmail.com

T. A. Mohammed
e-mail: tareq.mohammed@altinbas.edu.tr

H. H. Safi
College of Basic Education, Al-Mustansiriya University, Baghdad, Iraq

T. A. Mohammed
College of Science, Kirkuk University, Kirkuk, Iraq

Z. F. Al-Qubbanchi
Applied Science, University of Technology, Baghdad, Iraq
e-mail: Zenafawzy97@gmail.com

© Springer Nature Switzerland AG 2019
W. K. Wong (ed.), *Artificial Intelligence on Fashion and Textiless,*
Advances in Intelligent Systems and Computing 849,
https://doi.org/10.1007/978-3-319-99695-0_18

145

1 Introduction

In the scenic field the term of Multi-objective optimization (otherwise called multi-target programming, vector improvement, multi-criteria advancement, multi attribute enhancement or Pareto streamlining) is a range of various criteria basic leader-ship, that is worried with scientific improvement issues including more than one target capacity to be upgraded at the same time [1]. Multi-target streamlining has been connected in numerous fields of science, including building, financial aspects and logistics where ideal choices should be taken within the sight of exchange offs between two or all the more clashing destinations. Minimizing cost while expand-ing solace while purchasing an auto, and amplifying execution whilst minimizing fuel utilization and outflow of contaminations of a vehicle are case of multi-objective advancement issues including two and three destinations, separately. In down to earth issues, there can be more than three destinations [2]. For a nontrivial multi-objective enhancement issue, there does not exist a solitary arrangement that at the same time streamlines every goal. All things considered, the target capacities are said to strife, and there exists a (conceivably limitless) number of Pareto ideal arrangements [3]. An answer is called non-dominated, Pareto ideal, Pareto proficient or non-inferior, if none of the target capacities can be enhanced in worth without debasing a portion of the other target values. Without extra subjective inclination data, all Pareto ideal arrangements are considered similarly great (as vectors cannot be requested totally). Scientists study multi-target enhancement issues from various perspectives and, in this manner, there exist diverse arrangement rationalities and objectives when setting and explaining them [4]. The objective might be to locate a delegate set of Pareto ideal arrangements, and/or measure the exchange offs in fulfilling the diverse desti-nations, and/or finding a solitary arrangement that fulfills the subjective inclinations of a human decision (HD). There is many developmental algorithm has been made for multi-objective improvement issues, there has been a tepid enthusiasm for tack-ling dynamic multi-objective advancement applications. The test results depending the input of some scenarios and apply by using the to Minimize the Cost Function in Multiple, Objective the result by NSGA-II approach shows its efficacy in finding the best calculation for the cost function.

2 Multi-objective Optimization Using NSGA-II

2.1 The Multi-objective Problem

Generally, around the word there are numerous (potentially clashing) objective that should be streamlined at the same time. Under such circumstances there no more exists a solitary ideal arrangement yet rather an entire arrangement of conceivable arrangements of proportional quality. The field of Multi-Objective Optimization (MO) [5–7] manages synchronous advancement of various, potentially contending, objective functions. At the point when the starting steps of a project compacted, the compacting cost plan would be included until the wrapping up venture time, and this estimation of cash would be included for shorter time. This impact is clear for the cash that is included with deferring in exercises time, so the time esteem of cash is a powerful consider this region. In the exhibited model, we utilized the time estimation of cash to compute the best time for compacting or postponing the exercises. We added the compacting expense to the aggregate cost capacity and subtract the venue of sparing cash of postponing on the exercises from the aggregate cost capacity, so the Pert system improve tradeoff amongst time and cost, considering the time estimation of cash [8].

2.2 The Nature Problem of Multi-objective Optimization

Numerous choice circumstances call for basic leadership when more than one target should be considered (called multi-objective decision-making). Such a methodology is run of the mill to venture issues where business banks need to adjust return and hazard due to legitimate and moral commitments request [9].

Different issue where multi-objective enhancement fit the way of the issue being referred to are recorded next [10–12]:

- The outline of a mechanical part needs to meet a few objectives, for example, affectability amplification, unbending nature boost, and cost-minimization.
- A decision-making help is combined with the element NSGA-II systems to distinguish one arrangement from the got front consequently for the adjustment in the issue with time t can be either in its target capacities or in its limitation capacities or in its variable limits or in any blend of above.
- Building Electrical Power office is to be resolved in light of expanding electrical force however much as could reasonably be expected, wellbeing contemplations of the inhabitants, monetary states of the occupants, the economy of the city and the state, neighborhood governmental issues.

2.3 The Problem Frame

The main problem in economic environments is the high cost function project, for that there is need to proposed perfect solution for this issue. The trouble of recognizing an ideal solution for a multi-objective decision problem into decrease the cost function that may exist between the ideal answers for the different destinations [13, 14]. Indeed, even the meaning of "optimality" in this setting is not straight-forward. The best answer for a specific goal may be the most noticeably bad for another. The accompanying graphical in Fig. 1 presentation delineates the nearness of conceivable

Fig. 1 The accompanying
graphical for the conceivable
clash among targets

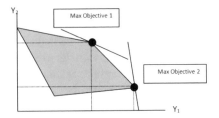

clash among targets. The objectives are to minimize direct cost operation to decrease
the cost of running operation.

3   The Using of NSGA-II

Non-dominated sorting genetic algorithm (NSGA-II) is one of the most proficient and
well known multi-target calculations, which was displayed by Deb [15] and Zade et al.
[16] and demonstrated its value in multi-target issues [13]. The NSGA-II can united
with Pareto sets of arrangements, and the outcomes could spread to all sets. NSGA-II
utilizes non-prevailing sorting for united certainty furthermore swarming separation
for cutting the terrible answers for winning better arrangements [14]. Absolutely, its
higher capability makes this calculation a decent choice for multi-objective problems.
At last, NSGA-II was chosen since it required less capacity assessments on a number
of compelled test issues, to the detriment of somewhat more terrible execution in
equitably spreading plans over the last Pareto front. This calculation is additionally
less demanding to adjust for taking care of limitations what's more, controlled elitism
taking into account proposals in Ref. [15]. The algorithm chooses arrangements
through a control sorting method. For this situation the best answer for get the
minimize in cost function is:

$$J_q = \sum_{j=1}^{n} \sum_{i=1}^{k} u_{ij}^q \cdot d^2(\vec{Z}_j, \vec{m}_i), \quad 1 \le q \le \infty, \tag{1}$$

where q is the fuzzy type, d demonstrates a separation measure between the j-th
design vector and i-th bunch centroid, and means the enrollment of j-th design in
the i-th group. The timer operation record is characterized as a component of the
proportion of the aggregate variety to the base partition sep of the groups. Here and
sep might be composed as:

$$\sigma = \sum_{i=1}^{k} \sum_{p=1}^{n} u_{ip}^2 \cdot d(\vec{m}_i, \vec{Z}_p) \tag{2}$$

and

$$\text{sep}(Z) = \min_{i \neq j}\{d^2(\vec{m}_i, \vec{m}_j)\} \qquad (3)$$

The timer operation (TO) written as:

$$\text{TO}_q = \frac{\sigma}{n \times \text{sep}(Z)} = \frac{\sum_{i=1}^{k} \sum_{p=1}^{n} u_{ip}^q \cdot d^2(\vec{m}_i, \vec{Z}_p)}{n \times \min_{i \neq j}\{d^2(\vec{Z}_i, \vec{Z}_j)\}} \qquad (4)$$

By giving the arrangement of focuses a chance to be indicated by the participation estimation of the j-th design in i-th group and are processed as:

$$u_{ij} = \frac{1}{\sum_{p=1}^{k} \left(\frac{d(\vec{m}_i, \vec{Z}_j)}{d(\vec{m}_p, \vec{Z}_j)}\right)^{\frac{2}{q-1}}}$$

Goal equations programming will find the solution where the total deviation from the goals is minimized.

4 The Proposed Solution

4.1 Material Image Database

At the point when no reasonable victor among the distinctive goals can be distinguished, tradeoffs among them ought to be considered. Building an added substance objective capacity where the distinctive destinations are weighted suitably, as per their relative significance, is one method for doing only that.

For the most part, a multi-objective choice issue with a few objective cost $Z1, Z2… Zk$ can be consolidated into one weighted whole target capacity. The determination of the relative weights (ri) will be talked about later (yet we have to remember this is a precarious and essential issue that may substantially affect the chose strategy). For the time being, we demonstrate that

- The added substance objective work that consolidates every one of the goals into a solitary target capacity is to be minimized.
- If a goal Zi is to be minimized, the added substance objective capacity incorporates the term $riZi$; and if Zi is to be expanded, the added substance objective capacity incorporates the term—$riZi$ (ri is certain).

Give us a chance to expect our multi- objective issue incorporates three goals $Z1$, $Z2, Z3$.

$Z1$ is minimized, $Z2$ and $Z3$ are boosted. The added substance target work that joins every one of the goals is Min $Z = r1Z1 - r2Z2 - r3Z3$.

Form a calculating cost function weighted-aggregate NSGA-II from the accompanying accumulations of target capacities:

(a) $\text{Min } 2M1 + 3M2 - 1M3$ (b) $\text{Min } 3M1 - 1M2$

$\qquad \text{Max } 4M1 - 2M2 \qquad\qquad \text{Min } 4T1 + 2T2 + 9T3$

$\qquad \text{Max } 1M2 + 1M3$

To calculate the best cost function for matter above we get the result below:

(a) $\text{Min } r1(2M1 + 3M2 - 1M3) - r2(4M1 - 2M2) - r3(1M2 + 1M3)$

$\quad = \text{Min}(2r1 - 4r2)M1 + (3r1 + 2r2 - r3)M2 - (r1 + r3)M3$

(b) $\text{Min } r1(3T1 - 1T2) + r2(4T1 + 2T2 + 9T3)$

$\quad = \text{Min}(3r1 + 4r2)T1 + (-r1 + 2r2)T2 + 9r2T3$

5 Experimental Results

The test results depending the input of some scenarios and apply by using the equation that we explain in previous paragraph, the viability of multi-objective by using NSGA-II based bunching has been accommodated six counterfeit and four genuine datasets. Table 1 shows the subtle elements of the datasets. The genuine datasets are iris, wine, bosom growth and the yeast sporulation information. The sporulation dataset is accessible from.

Table 1 The result of experiment

Input	NSGA-II	Font size and style
	M	Result
Input_1	9.37 (1.72)	0.669317 (0.0892)
Input_2	3.16 (0.072)	0.654393 (0.00927)
Input_3	3.57 (0.51)	0.765691 (0.005686)
Input_4	6.28 (0.46)	0.827618 (0.02871)
Input_5	12.43 (0.939)	0.768379 (0.005384)
Input_6	4.65 (1.58)	0.642091 (0.002833)

6 Conclusion and the Future Work

In this study, we introduced a framework that can minimize the high cost function in the economic environments by using the fast elitist multi-objective algorithm based on Non-dominated sorting genetic algorithm (NSGA-II). The proposed framework calculates the cost function weighted-aggregate NSGA-II from the accompanied accumulations of target capacities. The programming of objective equations can find the solution where the total deviation of the goals is minimized. The future work of this study can be applied to many complex real world problems especially for fashion optimization problems. In addition, it can be used to prove the efficiencsy of NSGA-II.

References

1. Shahriari, M.: Multi-objective optimization of discrete time—cost tradeoff problem in project networks using non-dominated sorting genetic algorithm. J. Ind. Eng. Int. **12**(2), 159–169 (2016). https://doi.org/10.1007/s40092s-016-0148-8
2. Yijie, S., Gongzhang, S.: Improved NSGA-II multi-objective genetic algorithm based on hybridization-encouraged mechanism. Chin. J. Aeronaut. **21**(6), 540–549 (2008). https://doi.org/10.1016/S1000-9361(08)60172-7
3. Bower, G. C., Kroo, I. M.: Multi-objective aircraft optimization for minimum cost and emissions over specific route networks, (X), 1–23 (n.d.)
4. Gadhvi, B., Savsani, V., Patel, V., Multi-objective optimization of vehicle passive suspension system using NSGA-II, SPEA2 and PESA-II. Procedia Technol. **23**, 361–368 (2016). http://doi.org/10.1016/j.protcy.2016.03.038
5. Shahriari, M., et al.: A new mathematical model for time cost trade-off problem with budget limitation based on time value of money. Appl. Math. Sci. **4**(63), 3107–3119 (2010)
6. Tiwari, S., Johari, S.: Project scheduling by integration of time cost trade-off and constrained resource scheduling. J Inst. Eng. (India) Ser. **96**(1), 37–46 (2015)
7. Deb, K., Pratap, A., Agarwal, S., Meyarivan, T.: A fast and elitist multiobjective genetic algorithm: NSGA-II. IEEE Trans. Evol. Comput. **6**, 182–197 (2002)
8. Gadhvi, B., Savsani, V. (eds.): Passive suspension optimization using teaching learning based optimization and genetic algorithm considering variable speed over a bump. In: ASME 2014 International Mechanical Engineering Congress and Exposition; 2014: American Society of Mechanical Engineers
9. Hays, J., Sandu, A., Sandu, C., Hong, D.: Parametric design optimization of uncertain ordinary differential equation systems. J. Mech. Des. **134**, 081003 (2012)
10. Das, R.K., Samal, C., Mallick, S.: Study of multi-objective optimization and its implementation using NSGA-II. Diss. (2007)
11. Deb, K., Member, A., Pratap, A., Agarwal, S., Meyarivan, T.: A fast and elitist multiobjective genetic algorithm **6**(2), 182–197 (2002)
12. Deb, K., Sundar, J., Udaya Bhaskara Rao, N.: Reference point based multi-objective optimization using evolutionary algorithms (n.d.)
13. Chase, N., Rademacher, M., Goodman, E., Averill, R., Sidhu, R.: A benchmark study of multi-objective optimization methods multi-objective optimization problem, 1–24 (n.d.)
14. Fonseca, C.M., Fleming, P.J.: Genetic algorithms for multiobjective optimization: formulation discussion and generalization. Icga. **93** (1993)

15. Deb, K., Karthik, S.: Dynamic multi-objective optimization and decision-making using modified NSGA-II: a case study on hydro-thermal power scheduling. In International conference on evolutionary multi-criterion optimization, pp. 803-817. Springer, Berlin, Heidelberg (2007)
16. Deb, K., Agrawal, S., Pratap, A., Meyarivan, T.: A fast elitist non-dominated sorting genetic algorithm for multi-objective optimization: NSGA-II (n.d.)

Two-Layer Mixture Network Ensemble for Apparel Attributes Classification

Tianqi Han, Zhihui Fu and Hongyu Li

Abstract Recognizing apparel attributes has recently drawn great interest in the computer vision community. Methods based on various deep neural networks have been proposed for image classification, which could be applied to apparel attributes recognition. An interesting problem raised is how to ensemble these methods to further improve the accuracy. In this paper, we propose a two-layer mixture framework for ensembling different networks. In the first layer of this framework, two types of ensemble learning methods, bagging and boosting, are separately applied. Different from traditional methods, our bagging process makes use of the whole training set, not random subsets, to train each model in the ensemble, where several differentiated deep networks are used to promote model variance. To avoid the bias of small-scale samples, the second layer only adopts bagging to mix the results obtained with bagging and boosting in the first layer. Experimental results demonstrate that the proposed mixture framework outperforms any individual network model or either independent ensemble method in apparel attributes classification.

Keywords Apparel attributes classification · Boosting · Ensemble · Bagging

1 Introduction

Automatically recognizing apparel attributes has recently drawn great interest in the computer vision community. The recognized apparel attributes can be used in various applications, for example, automatic product tagging, clothes searching, clothing style recognition and clothes matching strategy learning. However, the annotation

T. Han · Z. Fu · H. Li (✉)
AI Lab, ZhongAn Information Technology Service Co., Ltd., Shanghai, China
e-mail: lihongyu@zhongan.io

T. Han
e-mail: hantianqi@zhongan.io

Z. Fu
e-mail: fuzhihui@zhongan.io

© Springer Nature Switzerland AG 2019
W. K. Wong (ed.), *Artificial Intelligence on Fashion and Textiles*,
Advances in Intelligent Systems and Computing 849,
https://doi.org/10.1007/978-3-319-99695-0_19

of the clothes requires special fashion domain knowledge and careful data cleaning, so that the available datasets are quite limited, thus the trained model is leading to overfitting with a high probability. In the meanwhile, clothes, with high variation including deformation and occlusion, require that the recognition models have better capabilities of describing clothes features.

Most previous clothing recognition methods [1, 2] are based on the hand-designed features which are hardly optimal for the customized classification tasks. For deep learning methods, DeepFashion [3] was proposed recently to handle clothing deformations and occlusions via jointly predicting the landmark locations and clothing attributes which achieves robust performance on the clothing recognition problem. Methods based on various deep neural networks for image classification can also be applied to apparel attributes recognition. An interesting problem raised is how to ensemble these methods to further improve the accuracy of apparel attributes classification.

Ensemble can be regarded as a classification task based on the outputs from all the base predictors, which will mine the correlation among all the dimensions of one base predictor output and the inter-predictors' relationship. Generally speaking, ensemble methods can smooth the outputs of multiple classifiers and reduce the variances. Tree boosting techniques have demonstrated good performance in classification tasks. Recently, gradient tree boosting classifiers [4, 5] are proposed as a functional gradient descent problem which has given state of the art results on many classification problems.

In this paper, we propose a two-layer mixture framework for ensemble different networks. In the first layer of this framework, two types of ensemble learning methods, bagging and boosting, are separately applied. Different from traditional methods, our bagging process makes use of the whole training set, not random subsets, to train each model in the ensemble, where several differentiated deep networks are used to promote model variance [6]. To avoid the bias of small-scale samples, the second layer only adopts bagging to mix the results obtained in the first layer. Experimental results demonstrate that the proposed mixture framework performs better than both of the individual network model and independent ensemble method in apparel attributes classification.

2 Methodology

2.1 Ensemble Framework

The ensemble method is based on the assumption that different classifiers are complementary to each other. There are mainly two strategies to ensemble the predictors: bagging via straightforward weighted voting of the outputs, and boosting through training a new classifier by concatenating feature vectors from the predictors.

Fig. 1 Structure of the two-layer mixture network ensemble framework. In the first layer, base predictors are ensembled with bagging and boosting methods respectively. The second layer combines each model with the bagging strategy

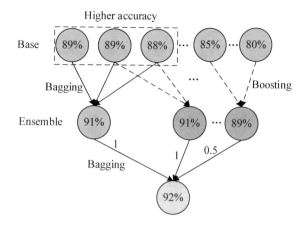

The bagging methods are effective for the small number of predictors and require no extra data. However, the ensemble results are highly correlated with the original predictors hence the accuracy cannot be consistently improved with the increase of the number of predictors. Boosting methods, such as XGBoost [5] and CatBoost [4], have been proposed to learn a new model for ensembling weak classifiers. After concatenating the output probabilities of weak classifiers to a vector as the input, a new predictor can be trained through boosting with a set of new training data, which is beneficial to avoiding the risk of overfitting.

Given the fixed amount of data, increasing the training data for ensemble will decrease the training data for CNN predictors. Empirical studies show that when the number of training data for ensemble is small, the performance of bagging and boosting is basically close. It is also observed that boosting produces different results from bagging even with the same predictors, which means that boosting the predictors may result in less correlation with bagging them. Based on this observation we propose a two-layer ensemble framework, as shown in Fig. 1. In the first layer, we pick out some base predictors to compose K groups (possibly overlapped) and use boosting methods to produce K new predictors. The new K predictors have relatively higher accuracy than base predictors after boosting. At the same time, some base predictors with high accuracy are bagged through weighted voting to ensure the robustness of the framework. Different from traditional methods, we trained differentiated deep networks to promote model variance. In this way, the whole training set, not random subsets are used in the bagging process. In the second layer, the bagging method is performed on these new $K + 1$ predictors. In the bagging process, the weight tuning is dependent on the capability of base predictors.

2.2 Mixture Network Ensemble

Resizing Strategy. For deep networks, input images generally have a fixed size during training. Directly upsampling original images probably changes the aspect ratio of images and deforms the objects in images. As a result, important clues for classification may be lost. For instance, some apparel attributes are with regards to the length of clothes. If the aspect ratio of a long skirt is changed in the process of resizing, the apparel attributes may be predicted as a mid-length skirt. In our resizing strategy, the aspect ratio of original images is fixed during scaling and the scaled images are padded with a certain RGB value for network training.

Predictor Grouping. Deep neural networks have shown great potentials in image recognition tasks. In the proposed ensemble framework, diverse deep networks are trained as the base predictors. The networks vary from the network architecture, the optimization method and the resolution of input images. In our implementation, 15 popular network models are utilized as base predictors for ensemble, including three Resnet50 [7] models with 256px, 384px and 512px resolution inputs (1:1 aspect ratio), one Resnet152 model with 512px inputs, one SE-Resnet50 [8] model with 512px inputs, one SE-Resnext50 model with 512px inputs, one SE-Resnext101 model with 384px inputs, two Inception-V4 [9] models with 299px and 512px inputs, two DenseNet121 [10] Models with 256px and 384px inputs and four DenseNet201 Models with 256px, 384px and 512px inputs. All the models are initialized by publicly available ImageNet [11] parameter values.

Data Augmentation. To improve the performance of base predictors, we augment original data through random flipping, rotation ($-45°$ to $+45°$), contrast adjustment (0.7–1.3), random Gaussian blur and so on.

3 Experiments

In this section, we evaluate the performance of single model and the ensemble methods on apparel attributes classification. All apparel images from the fashionAI competition are divided into 8 subsets each of which corresponds to an apparel attributes key. In each subset, there are 5000–20,000 images with annotated attributes. Apparel attributes classification aims to predict the attributes probability of unknown clothes images. Specifically, the fashionAI competition can be treated as eight independent and disjoint classification tasks. The overall classification accuracy is computed through averaging the accuracies of these eight classifiers. According to the protocol of the competition, the accuracy involving Top-1 prediction probabilities is described as the basic precision. In the process of training base predictors, 10% of original images are left out as validation and test data. During the boosting ensemble, the validation data act as the new training data. The fivefold cross-validation is conducted in our experiments to prevent overfitting as a result of the small dataset.

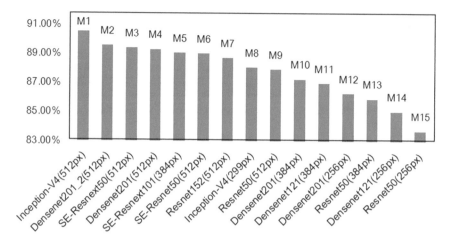

Fig. 2 Basic precision of 15 base predictors. The predictors are sorted and relabeled in accordance with the accuracy on the validation data

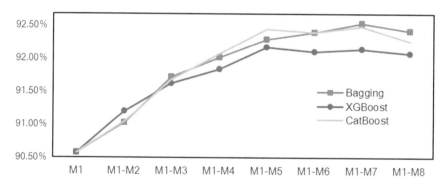

Fig. 3 Basic precision of three ensemble methods under different number of predictors

Figure 2 illustrates the basic precision of 15 base predictors on the validation data. It is observed that the image resolution networks can accept has a clear effect on the classification accuracy of apparel attributes. In addition, the well-designed network architecture, the reasonable optimization method and the deeper network can improve the accuracy.

For simplicity, our experiments adopt equal weights for base predictors in the bagging ensemble. Two methods, XGBoost and CatBoost, are used in boosting. Figure 3 illustrates the basic precision with three independent ensemble methods. It is worth noting that the accuracy keeps increasing when the number (N) of predictors is less than 5 ($N < 5$), but will vibrate with more predictors used. The best accuracy is 92.53%, where predictors M1–M7 are bagged.

We also computed the difference of predicted labels between the bagging on M1–M7 and other ensemble strategies on M1–M6 or M1–M5. The difference is

Table 1 Differences between predicted labels with bagging on M1–M7 and other ensemble strategies on M1–M6 or M1–M5

	M1–M6		M1–M5	
	Bagging (%)	Boosting (%)	Bagging (%)	Boosting (%)
Difference	1.23	**3.17**	1.37	**3.23**

The more difference is in bold

Table 2 Basic precision (BP) under different mixture strategies in the first layer

Mixture strategies	BP (%)
Bagging	92.53
Bagging + Boosting (M1–M6)	92.59
Bagging + Boosting (M1–M6) + Boosting (M1–M5)	92.69
Bagging + Boosting (M1–M6) + Boosting (M1–M5) + Boosting (M8–M12)	**92.76**

The highest accuracy is in bold

described as the ratio of the number of different labels to the number of all labels. As presented in Table 1, the boosting produces more differences than the bagging on M1–M6 or M1–M5, where the boosting result is the average of XGBoost and CatBoost. Experiments demonstrate that the results, with other ensemble strategies on M1–M6 or M1–M5, are somewhat different to those with the bagging on M1–M7, even if base predictors are approximate. It is the differentiation that indicates that mixing different strategies can further improve the classification accuracy.

As described in Table 2, several mixture strategies in the first layer are tested and compared on the basic precision. If only bagging (i.e., one-layer ensemble) is used, the basic precision is the lowest (92.53%) among them. Other two-layer mixture strategies usually perform better, as shown in the last 3 rows of Table 2. If more base predictors are included through boosting on M8–M12, the basic precision becomes the best (92.76%). This validates the effectiveness of the two-layer mixture network ensemble method in improving the accuracy of apparel attributes classification.

4 Conclusions

This paper presents a two-layer mixture framework for ensemble different networks. In the first layer of this framework, two types of ensemble learning methods, bagging and boosting, are separately applied. Different from traditional methods using random subsets, the bagging process makes use of the whole training set to train each model in the ensemble, where several differentiated deep networks are used to promote model variance [11]. To avoid the bias of small-scale samples, the second layer only adopts bagging to mix the results obtained with bagging and boosting in the first layer. Experimental results validates the effectiveness of the proposed mixture framework to improve the apparel attributes classification accuracy.

References

1. Yang, M., Yu, K.: Real-time clothing recognition in surveillance videos. In: IEEE International Conferences on Image Processing (2011)
2. Kalantidis, Y., Kennedy, L., Li, L.-J.: Getting the look: clothing recognition and segmentation for automatic product suggestions in everyday photos. In: ICMR (2013)
3. Luo, Z., Luo, P., Qiu, S., Wang, X., Tang, X.: DeepFashion: powering robust clothes recognition and retrieval with rich annotations. In: Proceeding of IEEE Conferences on Computer Vision and Pattern Recognition (2016)
4. Dorogush, A.V., Ershov, V., Gulin, A.: CatBoost: gradient boosting with categorical features support. In: Workshops on ML Systems at NIPS (2017)
5. Chen, T., Carlos, G.: XGBoost: a scalable tree boosting system. In: KDD (2016)
6. Breiman, L.: Bagging predictors. Mach. Learn. **24**(2), 123–140 (1996)
7. He, K., Zhang, X., Ren, S., Sun, J.: Deep residual learning for image recognition. In: Proceeding of IEEE Conferences on Computer Vision and Pattern Recognition (2016)
8. Hu, J., Shen, L., Sun, G.: Squeeze-and-excitation networks. In: arXiv preprint arXiv:1709.0 1507
9. Szegedy, C., Ioffe, S., Vanhouke, V., Alemi, A.A.: Inception-V4, inception-resnet and the impact of residual connections on learning. In: AAAI (2017)
10. Huang, G., Liu, Z., van der Maaten, L., Weinberger, K.: Densely connected convolutional networks. In: Proceeding of IEEE Conferences on Computer Vision and Pattern Recognition (2017)
11. Russakovsky, O., Deng, J., Su, H., Krause, J., Satheesh, S., Ma, S., Huang, Z., Karpathy, A., Khosla, A., Bernstein, M., Berg, A.C., Fei-Fei, L.: ImageNet large scale visual recognition challenge. In: IJCV (2015)

3D Digital Modeling and Design of Custom-Fit Functional Compression Garment

Rong Liu and Bo Xu

Abstract User-oriented custom-fit compression garment (CG) catering to individual preferences is fast growing. This study aims to establish and verify a new three-dimensional (3D) modeling and design system of custom-fit functional compression garment including digitalized body scanning, digital knitting design and performance assessment. The personalized numerical 3D models of human and elastic compression fabrics were created based on medical magnetic resonance scanning images, anthropometry, and finite element analysis. The developed numerical model was validated using experimental data and demonstrated agreement with the measured skin pressure profiles. The tissue deformations, curvature variations and internal stress distributions by the designed and fabricated CG were visualized via the developed numerical model and used for assessment of biomechanical effects of the designed CG on the specific subject. The applicability of the proposed digitalized development and assessment system for custom-fit pressure product was demonstrated, which would not only contribute to improving custom-fit, pressure function and wearing comfort, but also conform to the promising trend of automation and data exchange in modern textile and fashion manufacturing technologies.

Keywords 3D modeling · Custom-fit · Compression garment · Design Visualization

1 Introduction

Compression garment (CG) has been increasingly applied in health care, medical, sports, body beauty and fashion fields. The design of compression shell largely depends upon the materials employed and users' characteristics. Functional pressure leggings, tights, and body suits have been applied for years, to reduce muscle

R. Liu (✉) · B. Xu
Institute of Textiles and Clothing, The Hong Kong Polytechnic University,
Kowloon, Hong Kong
e-mail: rong.liu@polyu.edu.hk

© Springer Nature Switzerland AG 2019
W. K. Wong (ed.), *Artificial Intelligence on Fashion and Textiles*,
Advances in Intelligent Systems and Computing 849,
https://doi.org/10.1007/978-3-319-99695-0_20

oscillation, increase muscle oxygenation and speed up sports recovery [1]. Calibrated pressure levels (10–50 mmHg) with gradient patterns have been designed in socks, compression stockings, pantyhose and bandages for prophylaxis and treatment of chronic venous insufficiency [2]. Single or multilayered CG (30–80 mmHg) has been used to reduce torso or limb volume to increase lymph transport, decrease interstitial pressure and capillary filtration [3]. Integrating shaped-foams with pressure layers are used for liposuction in abdominal contouring process [4]. Elasticized compression shell with 6–50 mmHg are employed in treatment of hypertrophic burn scars as the standard first-line therapy [5, 6]. Despite the fact that the exact working mechanisms of action of compression has been not fully understood, and the optimal pressure for effective treatment has not been standardized in practice, various design and assessment approaches are fast developed to facilitate growth of CG in diverse applications. Effectively design and visualization of pressure function is highly demanded to raise users' satisfaction and to achieve the expected functional performances.

Different from the traditional CG, "tailored" pressure dosages are required in user-oriented functional compression wear. 3D modeling provides possibilities to scientifically explore pressure profiles to predict functional performances in an effective and intuitive manner, which contribute to providing an evidence-based solution to improve biomechanical effectiveness, ergonomic fit, and wearing comfort of CGs for the targeted consumers. This study aims to develop a digital system to promote effectiveness of CG design via a 3D Modeling technique integrating 3D digital body scanning, seamless digital knitting and biomechanical visualization. This digitalized system would be promising to enhance compression textiles design, and to facilitate pressure data display and data exchange in modern fashion and textiles production process.

2 Methods

2.1 3D Digital Body Scanning

The VITUS^{Bodyscan} scanner (Human Solutions of North America, Cary, NC) was used to collect digital anthropometric data of the studied subject. This scanning system produced a true-to-scale 3D model with precise and reproducible capture (point density: 200 points/cm^3, scan time: 6–10 s) when the subject was standing still with her feet placed at two standardized distance-apart footprints on the platform in the scanning booth. Table 1 shows the basic characteristics of the studied female subject.

Table 1 Basic characteristics of the studied subject in 3D modeling development

Age	Height (cm)	Weight (kg)	BMI(kg/m^2)	Condition	Pressure indicated
35	158.5	48.3	19.2	Healthy	15–21 mmHg for daily care
Circumference (cm)		Ankle: 21.7	Brachial: 28.3	Calf: 35.6	Knee: 33.9
Heights (cm)		Ankle: 10.8	Brachial: 19.1	Calf: 27.6	Knee: 35.2

2.2 3D Digital Design and Fabrication

Based on the anthropometric study, the designed and developed custom-made CG was a style of knee-high compression legging (CL) in length of 26 cm. According to standard RAL 387/1 and our extensive product development practice, the elongation of tubular fabrics in lengthwise and crosswise along the lower limb were up to 30% and 30–70%, respectively. The developed legging comprised three sequential segments to provide a degressive pressure gradient from ankle to calf (Fig. 1). The residual pressure ratios from ankle to calf were 100%, 100–70%, and 50–60%, respectively. The general pressure ranged from 15 to 21 mmHg to produce a degressive gradient pressure to counteract gravity force, and to reduce increased venous hypertension during standing still for prolonged period of time. Such legging can be used for healthcare, sports and body beauty purposes. Heterogeneous elastomers were interlaced by 1 × 1 laid-in knitting structures to form functional compression shell using Lonati L45ME knitting unit (cylinder diameter: 4.5 inches, 352 needles). The knitting materials included polyamide covered Lycra yarns (PCL) with linear density of 44dtex/44dtex as ground layer, and PCL yarns with linear density of 233 dtex/44dtex/as the laid-in thread. The mechanical properties of the developed fabric shells were determined as illustrated in Table 2.

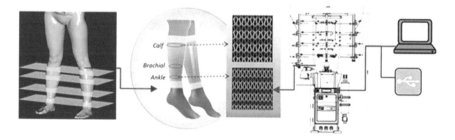

Fig. 1 From 3D body scanning, digital knitting design to 3D digital modeling

Table 2 Mechanical properties of the customized compression shell materials

Segments	W^a($\times$ 10–10 ton/mm^3)	$E1$ (MPa)	$E2$ (MPa)	$G12$ (MPa)	V	T (mm)
Ankle	3.25 ± 0.12	0.28 ± 0.02	0.19 ± 0.02	0.11 ± 0.01	0.27 ± 0.02	1.0 ± 0.1
Brachial	2.75 ± 0.11	0.21 ± 0.02	0.15 ± 0.01	0.08 ± 0.01	0.33 ± 0.03	0.9 ± 0.1
Calf	2.30 ± 0.10	0.15 ± 0.01	0.10 ± 0.01	0.05 ± 0.01	0.37 ± 0.03	0.8 ± 0.2

[a]W mass density; $E1$ and $E2$ Young's modulus in wale and course directions of fabric; $G12$ shear modulus; V Poisson's ratio; T thickness of fabric specimen

2.3 3D Digital Modeling

The bio-function of the custom-fit legging was digitally visualized via a developed 3D finite element model (FEM). To attain the anatomic structures of leg, multiple Magnetic Resonance Images (MRI) were taken with intervals of 2 mm in the neutral unloaded position. The boundaries of the skeletons and soft tissues were determined by employing MIMICS (V 7.10) (Materialise, Leuven, Belgium) software. The skin and soft tissue were assumed to be homogeneous, isotropic and linearly elastic biomaterials. The Young's modulus, Poisson's ratio, and mass density of soft tissues were taken as 0.02 Mpa, 0.48, and 1.03E-09 ton/mm^3, respectively. The bones including femur, tibia, fibula, and patella were taken as rigid and incompressible, and assumed to have no deformations under the pressure by legging. The boundary of the bones was set to zero displacements. The digital CL was divided into three portions (ankle, brachial, and calf) along the shaped cylinder. The numerical simulation of interaction between leg and fabric shell was performed within ANSYS/CAE (v.15.0) modeling environment with explicit approach (Fig. 2). The leg model was meshed with 10-node tetrahedron solid elements. The CL model was meshed with 4-node quadrilateral shell elements in a cone shape, which facilitated legging shell to slide onto the leg model easily in the simulation. The surface-to-surface contact was applied to simulate interface interactions between leg and CL. A specific displacement (100 mm) was set to edge of CL at pulling stage, where the contact friction coefficient was assumed to be zero. When the CL reached the position, the contact friction coefficient was changed into none zero using the function of the element death and birth. Applying this Modeling strategy, there was no displacement boundary on the legging at the end, where the legging was coupled with leg by frictional force and fabric tension. To intuitively reflect intervention effects of CL on skin surface structure, a curvature comb tool was used by CAD software Pro/ENGINEER (PTC INC, WildFire 5.0) function. Skin surface curvatures and tissue deformations at different cross-sectional leg were further visualized using the developed 3D Modeling. The simulated pressures were validated through comparison with the measured pressure data via a portable pneumatic pressure sensor system. The transducer linearity (R^2) was greater than 0.9999 and deviation was ±1 mmHg. The skin pressure was measured when the subject was in

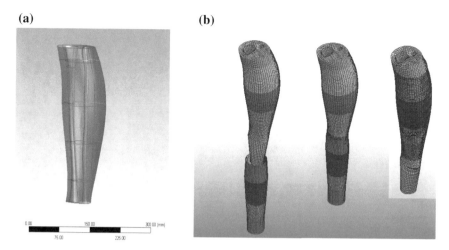

Fig. 2 **a** Sub-FE model of leg; **b** Simulated dynamic wearing process of CL

an upright standing position. Average pressure values were adopted based on three times of pressure measurements.

3 Results

A degressive gradient distributions from distal to proximal leg can be observed, which demonstrated a positive pressure function of CL to counteract gravity force and to reduce the increased venous hypertension in long-term standing (Fig. 3a). The external pressure further transmitted to the internal deeper tissues. Figure 3b illustrated heterogeneous internal stresses distributed at the major cross-sectional leg. The higher stresses were found at posterior muscular compartment and anterior tibia crest regions, as well as internal tibia and fibula bones with larger curvatures at local surfaces.

As shown in Fig. 4, the surface curvatures of leg cross-sections were uneven, which would result in uneven skin pressures beneath CL, especially for the bony ankle. The intervention of CL reduced fluctuations of curvature comb, indicating that the skin surface became flatten and smooth after application of compression garment.

The comparison on pressure values between the simulated and the measured data around the four key directions of lower limb was displayed in Fig. 5a–d. The measured pressure data were determined at the individual points; and the simulated data were adopted along the vertical path involving multiple continuous points of leg. More obvious degressive gradient pressure patterns were found at anterior and posterior leg compared with those at medial and lateral sides.

(a)

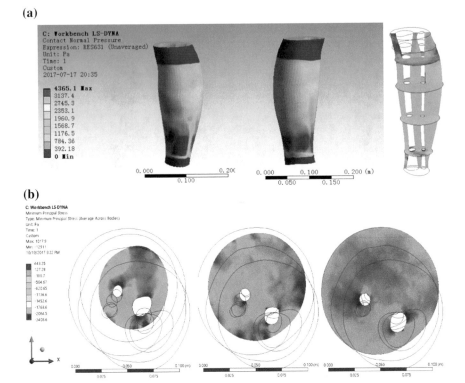

(b)

Fig. 3 **a** Longitudinal view of the simulated leg model; **b** Cross-sectional view of the simulated leg model

Further extracting the individual 16 points around four directions (anterior, posterior, lateral and medial) and along four heights (ankle, brachial, calf, knee) of the actual and the simulated lower limb, it can be seen that the simulated pressure values presented consistent trends with the measured ones as shown in Fig. 6, which demonstrated the applicability of the developed 3D modeling for visualizing and predicting custom-fit pressure function of the designed compression garment.

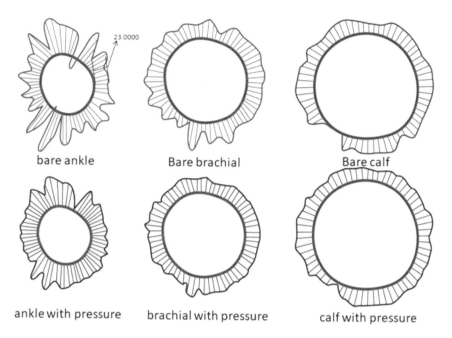

Fig. 4 Visualized curvature variations by intervention of CL using curvature comb tool

4 Conclusion

A digital-based design, fabrication and assessment system of compression garment has been developed via 3D seamless knitting technology and 3D biomechanical modeling based on digital anthropometric study. The developed 3D Modeling demonstrated applicability to act as an optional assessment approach to visualize and quantify pressure function of the engineered compression textiles. The way to intuitively illustrate the magnitudes and distributions of skin pressures, internal stresses, surface curvatures and tissue deformations of the subject's lower body by compression garments provide an evidence-based solution of pressure product design and development, which would improve effectiveness and custom-fit of functional compression garment in practical use; moreover, the digital-based knitting design, fabrication and 3D Modeling system will facilitate production automation and data exchange in modern fashion and textiles industries.

Acknowledgements We would like to thank the Hong Kong Polytechnic University for funding this research through research projects 1-ZE7K, 1-ZVLQ, G-YBUY and Innovation and Technology Fund through Project ITS/031/17.

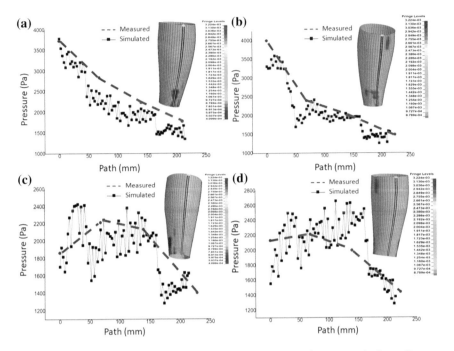

Fig. 5 **a** Pressure comparison on posterior leg; **b** Pressure comparsion on anterior leg; **c** Pressure comparison on lateral leg; and **d** Pressure comparison on medial leg

Fig. 6 Validation of the established 3D modeling on pressure function with experimental data

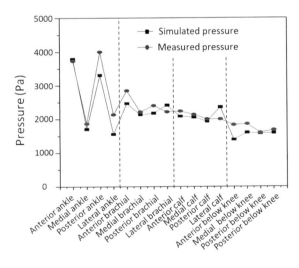

References

1. Leoz-Abaurrea, I., Nicholas, T., Aguado-Jiménez, R.: Heat dissipating upper body compression garment: thermoregulatory, cardiovascular, and perceptual responses. J. Sports Health Sci. (2016). https://doi.org/10.1016/j.jshs.2016.01.008
2. Liu, R., Guo, X., Lao, T.T., Little, T.J.: A critical review on compression textiles for compression therapy: textile-based compression interventions for chronic venous insufficiency. Text. Res. J. **87**(9), 1121–1141 (2017)
3. Slavin, S.A., Schook, C. C., Greene, A. K.: Lymphedema management. Support. Oncol. 211–220 (2011)
4. Hunstad, J.P., Repta, R.: Liposuction in abdominal contouring. Atlas Abdominoplasty 15–24 (2009)
5. Atiyeh, B.S., EI Khatib, A.M., Dibo, S.A.: Pressure garment therapy (PGT) of burn scars: evidence-based efficacy. Ann Burns Fire Disasters **26**(4), 205–212 (2013)
6. Macintyre, L.: New calibration method for I-scan sensors to enable the precise measurement of pressures delivered by 'pressure garments'. Burns **37**, 1174–1181 (2011)

Fine-Grained Apparel Image Recognition Based on Deep Learning

Jia He, Xi Jia, Junli Li, Shiqi Yu and Linlin Shen

Abstract There are many styles and details of apparel, including coat length, collar design, sleeve length and other detail properties. The e-commerce platform that manages apparel products needs to quickly and effectively identify the attribute categories of apparel for quick retrieval. Apparel image data contains many detailed features that can be easily deformed and occluded. Traditional image recognition technology has been unable to meet the requirements of its classification accuracy. The neural network based on deep learning can classify the fine-grained attributes of complex objects well after training. In this work, we use the apparel image data to train convolutional neural network for the classification of fine-grained attributes. To improve the classification accuracy, we also integrate the results of different models. The experiments show that the results of multi-model fusion are better than those of single model.

Keywords Fine-grained image recognition · Deep learning · CNN

1 Introduction

Currently, the main methods of commodity retrieval on e-commerce platform are text keyword and commodity label based. For text keywords search, the keywords entered by the user not necessarily reflect their real requirements. While category information could help users to refine the search scope, this approach requires a highly structured backend database. When the product label information is not clear, it will be difficult to find unknown brand cosmetics and apparel, etc. Another new way to retrieve goods is "photo shopping", which searches for the target goods in the e-commerce database by uploading the photo. Compared with the previous two

J. He · X. Jia · J. Li · S. Yu · L. Shen (✉)
Computer Vision Institute, School of Computer Science and Software Engineering,
Shenzhen University, Shenzhen, China
e-mail: llshen@szu.edu.cn

© Springer Nature Switzerland AG 2019
W. K. Wong (ed.), *Artificial Intelligence on Fashion and Textiles*,
Advances in Intelligent Systems and Computing 849,
https://doi.org/10.1007/978-3-319-99695-0_21

171

retrieval methods, the method has the advantage that the required goods can be retrieved without any text description.

As one of the major categories of e-commerce products, apparel is characterized by various styles and details. When it is difficult to quantify them by text keywords, we can search the related items by images using the technology of apparel image recognition. Apparel images have a variety of attribute dimensions such as coat length, collar design, lapel design and so on. Each attribute has a different category. For example, the category of collar design includes shire collar, Peter pan, pure collar and others. As there are many kinds of apparel products, and the design of apparel contains a lot of detailed features, it is very difficult to distinguish them. Furthermore, apparel is a flexible object and easy to be deformed or obscured and subjected to variance of lighting conditions.

In this paper, we use convolution neural network to achieve fine-grained classification of different attribute of apparel products, which can be applied to apparel retrieval, video shopping, clothing matching recommendation, automatic apparel labeling and so on.

Image classification is an important research topic in the field of computer vision. Fine-grayed classification aims to identify a more detailed subclass of the coarse-grained category, i.e., identifying different categories of flowers [1], birds [2], dogs [3], cars [4], etc. Fine-grained classification is more difficult than common image classification tasks because of the large intra-class differences and inter-class similarities among categories. Traditional classification algorithms based on hand-crafted features usually extract SIFT [5] or HOG [6] features from images. These features are limited in description capability and have poor recognition performance for fine-grained classification problems. In recent years, convolutional neural networks based on deep learning have achieved great success in the field of computer vision [7, 8]. Convolutional neural networks have also been applied to fine-grained image classification tasks, and great progress has been made.

Fine-grained image classification problem can be divided into two categories, i.e., strong supervision and weak supervision, according to the amount of supervision information used in training. Fine-grained classification based on strong supervision information means that in model training, in addition to the category label, the bounding-box and location of objects are also required. For example, Part-based RCNN [9] uses RCNN [10] algorithm to detect the location and bounding-box of object for fine-grained classification. Pose Normalized CNN [11] applied posture alignment to the image regions, and extracts convolutional features of different layers for feature fusion.

Fine-grained classification based on weak surveillance information only relies on image-level category labels for classification. Two Level Attention Model [12] fused feature information at both object and part levels. Constellation [13] uses convolution network to locate key points, and then uses these key points to extract local region information. Bilinear CNN [14] is an end-to-end network model that integrates feature extraction and model training into a single pipe line.

Table 1 The details of the dataset

AttrKey	AttrValues number	Train	Test
Coat length	8	11320	1453
Collar design	5	8393	1082
Lapel design	5	7034	900
Neck design	5	5696	708
Neckline design	10	17148	2095
Pant length	6	7460	949
Skirt length	6	19333	1153
Sleeve length	9	13299	1740

2 Dataset

The dataset used in this study was provided by 2018 FashionAI Global Challenge—Attributes Recognition of Apparel [15]. The organizers collected apparel images from Alibaba e-Commerce platform and labeled them with eight attribute dimensions, namely neckline design, collar design, high neck design, lapel design, sleeves length, length of top, length of skirt, and length of trousers. Each attribute has 5–10 categories. As shown in Table 1, The dataset contains about 90,000 training images and about 10,000 test images. The images have a resolution of 500×500 pixels and are stored in JPG file format. The size of each image is about 100 KB.

3 Method

We treat the classification of eight attribute dimensions as eight independent image classification tasks, and train a model for each attribute dimension separately. Because the dataset provided by the competition is not large enough, we use the model trained on ImageNet dataset for fine-tuning. ImageNet has more than 14 million images and 1000 categories, and models trained on this dataset already have good image recognition capabilities.

3.1 Data Preparation

We divide the dataset as training and validation set according to the ratio of 9: 1 in our experiment, 90% for training and 10% for validation. During the training stage, we generate additional training instances by random crops and mirroring. And the brightness, contrast and saturation of the picture are randomly adjusted. In this work, the original image is resized into 512×512, and randomly cropped into 448×448.

3.2 Network Structure

We chose ResNet [16] and DenseNet [17] as the networks for attribute classification in this paper. Both of them achieved very competitive results in the ImageNet challenges. Specifically, the DenseNet is composed by a number of dense blocks. It has the dense connections between each convolutional layer and all subsequent layers.

As shown in Fig. 1. The network employed DenseNet as base net. Since the number of training data is not large enough, even after data augmentation, we employed the weights trained on ImageNet as initialization. The Apparel images were employed for network fine-tuning. Using a pre-trained network as initialization, we can build a good base for feature extraction. The requirement of the number of training images can also be reduced.

We redefine the output layer according to the requirements of this task and initialize the weights randomly. We use the SGD algorithm to optimize the network.

The output of the network is normalized by softmax function to a legal probability value representing the probability that the sample belongs to each category. Softmax function can be defined as following:

$$p(x_i) = \frac{\exp(x_i)}{\sum_{j=1}^{n} \exp(x_j)} \tag{1}$$

We choose Cross-Entropy as loss function for classification tasks, that is defined as

$$H(p,q) = \sum_i p_i \times \log\left(\frac{1}{q_i}\right) = -\sum_i p_i \times \log(q_i) \tag{2}$$

In the formula above, the discrete p_i is the true label, and the given distribution q_i is the predicted value of the current model.

3.3 Network Training

This experiment is conducted on a Linux operating system server, using multiple NVIDIA Tesla P100 (16 GB memory) GPUs for accelerated training. The proposed

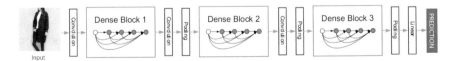

Fig. 1 The structure of single classification network

framework is based on MXNet and its python interface Gluon was used to train and test the network.

The eight networks in the proposed method were trained independently. We used different learning rate to train at different stages. The learning rate for the convolutional layers and fully connected layers is set to 0.001 at the first stage (10 epochs) and decreased 10 times in every 10 epochs. We trained the network with a total number of 40 epochs.

3.4 Network Prediction

When making the prediction, the original Apparel image is resized to 512×512, and cropped into 448×448 with reference to the center of the image. All the test images are sent to neural network for forward propagation prediction, and the probability for each attribute category is obtained. Then, the output results of the models are stored in a CSV file.

In order to improve the accuracy of the results, we select a few good models from the trained ones, and then fuse and integrate their prediction results. For the predicted values of each sample, we compare the probability values of the predicted results of different models and select the one with maximum probability. The prediction of larger probability values is employed as the final result of the image. For the T models we obtained $g_1, g_2, \ldots g_T$, we mix them into a new model G_x by the following formula:

$$G(x) = \arg \max_{t \in 1,2,..T} P_{rob}(g_t) \tag{3}$$

4 Experimental Result

4.1 Evaluation Criteria

In this work, the evaluation process is listed as follows:

(1) For each testing sample and each of the eight attribute dimensions, find out the maximum probability (MaxAttrValueProb) and its corresponding attribute value (MaxAttrValue).
(2) Given a preset threshold, ProbThreshold, for all the predictions whose MaxAttrValueProb exceeds ProbThreshold, count the number of samples whose attributes are correctly predicted and calculate Precision as the ratio to the total number of samples.
(3) Calculate the mean of all Precisions for different thresholds and record it as AP. In particular, the mean Precision of the eight attribute dimensions, when

ProbThreshold = 0, is recorded as BasicPrecision for model performance evaluation.

(4) mAP, the mean of APs of eight attribute dimensions, is used as another criterion for model performance evaluation.

4.2 Results

Table 2 lists the BasicPrecision and mAP of the DenseNet-161 network on the validation set, for all of the eight attribute dimensions. The network achieved the best performance, i.e., 0.98 mAP, for the attribute of skirt length, which provide the largest number of training images. One can observe from the two tables that the performance of the network is correlated with the number of available training images, i.e., the more training data available, the better validation precision can be achieved.

We also show the performance of ResNet with different layers in Table 3, together with that of DenseNet and their fusions. It can be seen that deeper networks perform better than shallower ones. For the same network, increasing the resolution of training pictures sent into the network can effectively improve the performance of the model. For example, Resnet152 trained using images with 512 achieved better performance than that trained using images with 256. Compared with a single model, the fusion of Resnet and DenseNet further improved the mAP from 0.9772 to 0.9785.

5 Conclusion

In this work, we proposed an approach to fuse ResNet and DenseNet to classify the fine-grained attributes of apparel images. The network trained on ImageNet dataset is used as a pre-training model and fine-tuned using the apparel images provided by FashionAI competition.

Table 2 The results of DenseNet-161 for different attribute dimensions

AttrKey	Image size	BasicPrecision	mAP
Coat length	512	0.835	0.907
Collar design	512	0.861	0.92
Lapel design	512	0.919	0.955
Neck design	512	0.852	0.917
Neckline design	512	0.862	0.918
Pant length	512	0.867	0.929
Skirt length	512	0.965	0.98
Sleeve length	512	0.908	0.951

Table 3 The results of different models

Model	Image size	BasicPrecision	mAP
Resnet50	256	0.8505	0.9532
Resnet152	256	0.8800	0.9644
Resnet152	512	0.9111	0.9756
Densenet161	512	0.9108	0.9772
Resnet152 + Densenet161	512, 512	0.9202	0.9785

From the experiments, we find that the deep convolution neural network can effectively identify the attribute categories of apparel images. The performance of the network for different attribute dimensions is correlated with the number of available training images, i.e., the more training data available, the better validation accuracy can be achieved. Compared with a single model, the fusion of ResNet and DenseNet further improved the mAP from 0.9772 to 0.9785, which is quite promising.

References

1. Nilsback M.E., Zisserman A.: Automated flower classification over a large number of classes. In: Sixth Indian Conference on Computer Vision, Graphics and Image Processing, ICVGIP'08, pp. 722–729 (2008)
2. Wah C., Branson, S., Welinder, P., Perona, P., Belongie, S: The caltech-ucsd birds-200-2011 dataset, (2011)
3. Khosla A., Jayadevaprakash N., Yao B., Li, F. F.: Novel dataset for fine-grained image categorization: Stanford dogs. In Proceedings of CVPR Workshop on Fine-Grained Visual Categorization (FGVC), vol. 2, p. 1 (2011)
4. Krause J., Stark M., Deng J., Li F.F.: 3d object representations for fine-grained categorization. In: IEEE International Conference on Computer Vision Workshops (ICCVW), pp. 554–561 (2013)
5. Lowe D.G.: Object recognition from local scale-invariant features. In: Proceedings of the Seventh IEEE International Conference on Computer Vision, vol. 2, pp. 1150–1157 (1999)
6. Dalal N., Triggs B.: Histograms of oriented gradients for human detection. In: IEEE Computer Society Conference on Computer Vision and Pattern Recognition, CVPR 2005, vol. 1, pp. 886–893 (2005)
7. Krizhevsky A., Sutskever I., Hinton, G.E.: Imagenet classification with deep convolutional neural networks. In: Advances in Neural Information Processing Systems, pp. 1097–1105 (2012)
8. Goodfellow I., Pouget-Abadie J., Mirza M., Xu B., Warde-Farley D., Ozair, S., Bengio, Y.: Generative adversarial nets. In Advances in Neural Information Processing Systems, pp. 2672–2680 (2014)
9. Zhang N., Donahue J., Girshick R., Darrell T.: Part-based R-CNNs for fine-grained category detection. In European Conference on Computer Vision, Springer, Cham, pp. 834–849 (2014)
10. Girshick R., Donahue J., Darrell T., Malik J.: Rich feature hierarchies for accurate object detection and semantic segmentation. In Proceedings of the IEEE Conference on Computer Vision and Pattern Recognition, pp. 580–587 (2014)
11. Branson S., Beijbom O., Belongie S.: Efficient large-scale structured learning. In: IEEE Conference on Computer Vision and Pattern Recognition, CVPR 2013, pp. 1806–1813 (2013)

12. Xiao T., Xu Y., Yang K., Zhang J., Peng Y., Zhang Z.: The application of two-level attention models in deep convolutional neural network for fine-grained image classification. In: IEEE Conference on Computer Vision and Pattern Recognition, CVPR 2015, pp. 842–850 (2015)
13. Simon M., Rodner E.: Neural activation constellations: unsupervised part model discovery with convolutional networks. In: Proceedings of the IEEE International Conference on Computer Vision, pp. 1143–1151 (2015)
14. Lin T.Y., RoyChowdhury A., Maji S.: Bilinear cnn models for fine-grained visual recognition. In Proceedings of the IEEE International Conference on Computer Vision, pp. 1449–1457 (2015)
15. FashionAI Global Challenge-Attributes Recognition of Apparel. https://tianchi.aliyun.com/competition/information.htm?spm=5176.100067.5678.2.55ab3a26kPTZlN&raceId=231649 (2018)
16. He K., Zhang X., Ren S., Sun J.: Deep residual learning for image recognition. In Proceedings of the IEEE Conference on Computer Vision and Pattern Recognition, CVPR 2016, pp. 770–778 (2016)
17. Huang G., Liu Z., Weinberger K.Q., van der Maaten, L.: Densely connected convolutional networks. In Proceedings of the IEEE Conference on Computer Vision and Pattern Recognition, pp. 2261–2269 (2017)

Learning a Discriminative Projection and Representation for Image Classification

Zuofeng Zhong, Jiajun Wen, Can Gao and Jie Zhou

Abstract Image classification is a challenging issue in pattern recognition due to complex interior structure for data. Meanwhile, the high-dimension data leads to heavy computational burden. To overcome these shortcomings, in this paper, we learn a discriminative projection and representation in a unified framework for image classification task. This method seeks a discriminative representation in a low-dimension space for an image, which enhances the classification accuracy and efficiency. Thus, a projection matrix is learnt by a criterion which demands the minimum of within-class residual and maximum of between-class residual in an iterative procedure. Then all samples are projected into a low-dimension space, and obtain the discriminative representation via L2 regularization. The experimental results demonstrate that the proposed method achieves better classification performances, compared with state-of-the-art sparse representation methods.

Keywords Image classification · Discriminative projection
Discriminative representation

Z. Zhong
Harbin Institute of Technology, Shenzhen Graduate School, Shenzhen 518055, China

J. Wen (✉) · C. Gao · J. Zhou
College of Computer Science and Software Engineering,
Shenzhen University, Shenzhen 518055, China
e-mail: enjoy_world@163.com

J. Wen · C. Gao · J. Zhou
Institute of Textiles and Clothing, Hong Kong Polytechnic University,
Kowloon, Hong Kong

J. Wen
The Hong Kong Polytechnic University Shenzhen Research Institute, Shenzhen 518055, China

© Springer Nature Switzerland AG 2019
W. K. Wong (ed.), *Artificial Intelligence on Fashion and Textiles*,
Advances in Intelligent Systems and Computing 849,
https://doi.org/10.1007/978-3-319-99695-0_22

1 Introduction

Classifying an image from a mass of image data is a challenging issue in pattern recognition [1]. Recently, numerous classification methods were designed to obtain superior classification performance. However, satisfactory classification result is still a pursuing target due to complex and diverse interior structure of image data. The study of discriminative representation suggests a new path to solve the classification problem based on the improvements of effectiveness. Recently, these methods obtained attractive performance in different recognition tasks, including face recognition [2], gait recognition [3], and action recognition [4].

Sparse representation classification (SRC) [2], as a typical representation-based classifier, presented a superior ability to solve the problem of face recognition on different conditions. SRC mainly employed the reconstruction residuals from a linear combination of training samples in all the classes to classify a sample. The representation coefficients indicate sparse solutions which are obtained by l_1 regularization. However, the l_1 regularization always needs an iterative process to obtain a numerical solution. Thus, the collaborative representation classification (CRC) [5], a l_2 regularization-based representation, is proposed to enhance the computational efficiency. But the robustness is still a suffering problem for classifying the images in real-world applications.

2 The Proposed Method

To seek the effective representation in a low-dimension space for images, the proposed method learns a discriminative projection and representation in a framework. This method can project the sample into a new low-dimension space which has minimum within-class representation residual and maximum between-class residual simultaneously. Then the projected sample is produced a discriminative representation via l_2 regularization. Therefore, we need to exploit the projection and representation matrices in this framework for image classification task. Previously, we introduce the discriminative representation-based classifier. In this classification framework, we learn a projection matrix based on a criterion which has maximum between-class differences and minimum within-class differences.

2.1 Discriminative Representation-Based Classifier

Let set $X = \{x_i | x_i \in \mathbb{R}^m, i = 1, \ldots, n\}$ as a training sample set, which has n training samples. If there are c classes and each class has s samples, X can be denoted as matrix $X = [X_1, \ldots, X_j, \ldots, X_c] = [x_1, \ldots, x_{s(j-1)+1}, \ldots, x_{sj}, \ldots, x_n], j = 1, \ldots, c$. vector y is denoted as test sample. Then, we can present the discriminative representation-based classifier. The model in Eq. (1) is written as the following formula:

$$\min_{Q}\|y - XQ\|_2^2 + \alpha \sum_{i=1}^{c} \|X_i Q_i\|_2^2 \tag{1}$$

Because of the convexity and differentiability of Eq. (1), we can obtain the optimal solution by taking the derivative with respect to Q and setting it to 0. The computational procedure is presented as follows:

Let $f(Q) = \|y - XQ\|_2^2 + \alpha \sum_{i=1}^{c} \|X_i Q_i\|_2^2$. The derivative with respect to Q of the first term of $f(Q)$ is

$$\frac{d}{dQ}\|y - XQ\|^2 = -2X^T(y - XQ). \tag{2}$$

Then we need to determine the derivative of the second term $\frac{d}{dQ}\left(\alpha \sum_{i=1}^{c} \|X_i Q_i\|_2^2\right)$.

Because $g(Q) = \alpha \sum_{i=1}^{c} \|X_i Q_i\|_2^2$ does not explicitly contain Q, it needs to compute partial derivative $\frac{dg}{dQ_k}$, and combine all $\frac{dg}{dQ_k}$ $(k = 1, \ldots, c)$ to achieve $\frac{dg}{dQ}$.

$g(Q)$ is composed of c terms which are dependent of Q_k. It firstly calculates the c partial derivatives $\frac{dg}{dQ_k}$ as follows:

$$\frac{\partial g}{\partial Q_k} = \frac{\partial}{\partial Q_k}\left(\alpha \sum_{i=1}^{c} \|X_i Q_i\|_2^2\right) = \alpha \sum_{i=1}^{c} \frac{\partial}{\partial Q_k}\|X_k Q_k\|_2^2$$

$$= \alpha c \frac{\partial}{\partial Q_k}\|X_k Q_k\|_2^2 = 2\alpha c X_k^T(X_k Q_k). \tag{3}$$

Thus the derivative $\frac{dg}{dQ}$ is

$$\frac{dg}{dQ} = \begin{pmatrix} \frac{\partial g}{\partial Q_1} \\ \vdots \\ \frac{\partial g}{\partial Q_c} \end{pmatrix} = \begin{pmatrix} 2\alpha c X_1^T(X_1 Q_1) \\ \vdots \\ 2\alpha c X_c^T(X_c Q_c) \end{pmatrix} = 2\alpha c \begin{pmatrix} X_1^T X_1 & \cdots & O \\ \vdots & \ddots & \vdots \\ O & \cdots & X_c^T X_c \end{pmatrix} Q. \tag{4}$$

Let $M = \begin{pmatrix} X_1^T X_1 & \cdots & O \\ \vdots & \ddots & \vdots \\ O & \cdots & X_c^T X_c \end{pmatrix}$, the derivative of the second term is $\frac{dg}{dQ} = 2\alpha c M Q$.

Combining the derivatives of the first and second terms, the derivative of Eq. (1) is $\frac{df}{dB} = -2X^T(y - XQ) + 2\alpha c M Q$. Thus, setting the derivative to zero, we have

$$2X^T X Q - 2X^T v + 2\alpha c M Q = 0, \tag{5}$$

which leads to

$$Q = (X^T X + \alpha c M)^{-1} X^T y \tag{6}$$

As a result, optimal Q is a closed-form solution.

Finally, we use Eq. (6) to compute representation coefficient Q of the test sample y. Then, a test sample v is classified to the kth class according to the following procedure,

$$k = \arg \min_i \| v - F_i Q_i \|_2^2 \tag{7}$$

2.2 Projection Learning

In the above the classification framework, we can learn a projection matrix to map the sample into a low-dimension space $\{a | a \in \mathbb{R}^d, (d < m)\}$. Inspired by [6], we use a decision criterion to obtain the projection matrix. This criterion attempts to make the representation residual of within-class samples as small as possibility and the representation residual of between-class samples as large as possibility, which is presented as follows:

$$J(P) = \frac{tr(P^T S_b^r P)}{tr(P^T S_w^r P)}, \tag{8}$$

where P is the projection matrix, and S_b and S_w are between-class and within-class scatter matrices respectively. These two matrices are calculated by where $S_b^r = \frac{1}{n(c-1)} \sum_{i,j} \sum_{s \neq i} [x_{ij} - r_{ij}][x_{ij} - r_{ij}]^T$ and $S_w^r = \frac{1}{n} \sum_{i,j} [x_{ij} - r_{ij}][x_{ij} - r_{ij}]^T$ respectively, where r_{ij} is the representation result of sample x_{ij}. Thus, we implement the singular value decomposition (SVD) to $S_b^r S_w^{rT}$ for obtaining an orthogonal matrix. Thus, orthogonal matrix P maps the samples into the space spanned by the d largest eigenvectors. Finally, we set above procedure as a iterative step to obtain the optimal projection matrix.

3 Experiments

To evaluate the performance of the proposed method, in this section, we conducted the experiments on several datasets; including the face dataset CMU Multi-PIE [12] and object image dataset Coil-100 [13]. In addition, state-of-the-art sparse representation

Table 1 Classification accuracies of different methods on the CMU Multi-PIE dataset

Number of training samples per subject	4	6	8	10
The proposed method	**93.42**	**99.08**	**99.81**	**99.85**
CRC	90.13	96.63	98.86	99.46
L1LS	92.79	**99.08**	99.66	99.81
FISTA	46.74	a50.01	57.26	71.16
Homotopy	55.49	56.55	63.03	71.25
DALM	68.05	67.62	72.27	84.67
INNC	89.80	95.09	98.01	99.31

The results highlighted with bold type are the best classification accuracies among different methods

methods were also test in our experiments. They are CRC [5], L1LS [8], FISTA [11], Homotopy [9], Dual augmented lagrangian method (DALM) [10], and LDA [7].

3.1 Experiments on the Face Recognition

This experiment tests the proposed method in face recognition task on the CMU Multi-PIE face dataset. A subset composed of 249 persons under 20 different illumination conditions with a frontal pose and seven different illumination conditions with a smile expression was selected. We choose face images corresponding to the first 3, 5, 7, 9 illuminations from the 20 illuminations and only one image from seven smiling images as training samples and use remaining images as testing samples. Parameter α is set to 0.00001 and the dimension is 250. In this experiment, the best results will be highlighted with bold type. Table 1 lists the recognition accuracy on four testing sets obtained using different methods. From the results, it is observed that our method obtains better classification accuracy than other methods.

3.2 Experiments on the Object Classification

Meanwhile, the object classification experiment was also conducted to test the proposed method. The used Coil-100 dataset consists of 7200 images taken from 100 classes and each class has 72 images. The first 10, 20, 30, and 40 images of each subject were used as training samples and the others were treated as the test samples, respectively. Parameter α was set to 1. Table 2 shows the classification accuracies of the proposed method and compared methods on this dataset. From the results, the

Table 2 Classification accuracies of different methods on the Coil-100 dataset

Number of training samples per subject	10	20	30	40
The proposed method	**55.21**	**63.56**	**78.51**	**80.69**
CRC	43.57	49.64	57.24	60.54
L1LS	48.89	56.92	68.57	74.34
FISTA	49.32	53.59	65.93	70.81
Homotopy	51.34	60.44	73.90	74.63
DALM	54.76	63.60	77.55	79.13
LDA	45.48	54.40	67.29	75.47

The results highlighted with bold type are the best classification accuracies among different methods

classification accuracies of the proposed method are all dramatically improved under different conditions.

4 Conclusion

To obtain an effective representation in a low-dimension space, we learn a discriminative projection and representation in the unified framework for image classification task. This method obtains superior classification performance by a discriminative representation in a low-dimension space for each image. Therefore, the classification framework simultaneously learns a projection matrix in an iterative procedure and produces a discriminative representation. In addition, the proposed method provides a computational efficient algorithm for image classification tasks.

Acknowledgements This work was supported in part by the Natural Science Foundation of China under Grant 61703283, 61773328, 61672358, 61703169, 61573248, in part by the research grant of the Hong Kong Polytechnic University (Project Code: G-UA2B), in part by the China Postdoctoral Science Foundation under Project 2016M590812, Project 2017T100645 and Project 2017M612736, in part by the Guangdong Natural Science Foundation under Project 2017A030310067, Project with the title Rough Sets-Based Knowledge Discovery for Hybrid Labeled Data and Project with the title The Study on Knowledge Discovery and Uncertain Reasoning in Multi-Valued Decisions.

References

1. Zhang Z., Xu Y., Shao L., Yang J.: Discriminative block-diagonal representational learning for image recognition. IEEE Transactions on Neural Networks and Learning systems, https://doi.org/10.1109/tnnls.2017.2712801 (2017)

2. Wright, J., Yang, A.Y., Ganesh, A., Sastry, S.S., Ma, Y.: Robust face recognition via sparse representation. IEEE Trans. Pattern Anal. Mach. Intell. **31**(2), 210–227 (2009)
3. Lai, Z., Xu, Y., Jin, Z., Zhang, D.: Human gait recognition via sparse discriminant projection learning. IEEE Trans. Circuits Syst. Video Technol. **24**(10), 1651–1662 (2014)
4. Wen, J., Lai, Z., Zhan, Y., Cui, J.: The L2,1-norm-based unsupervised optimal feature selection with applications to action recognition. Pattern Recogn. **60**, 515–530 (2016)
5. Zhang L., Yang M., Feng X.: Sparse representation or collaborative representation: Which helps face recognition? In: Proceedings of the International Conference on Computer Vision (ICCV), pp. 471–478 (2011)
6. Yang, J., Chu, D., Zhang, L., Xu, Y., Yang, J.Y.: Sparse representation classifier steered discriminative projection with applications to face recognition. IEEE Trans. Neural Netw. Learn. Syst. **24**(7), 1023–1035 (2013)
7. Belhumeur, P.N., Hespanha, J.P., Kriegman, D.: Eigenfaces vs. fisherfaces: recognition using class specific linear projection. IEEE Trans. Pattern Anal. Mach. Intell. **19**(7), 711–720 (1997)
8. Kim, S.J., Koh, K., Lustig, M., Boyd, S., Gorinevsky, D.: An interior-point method for large-scale l 1-regularized least squares. IEEE J. Sel. Top. Sig. Process. **1**(4), 606–617 (2007)
9. Yang A.Y., Ganesh A., Sastry S.S., Ma, Y.: Fast 1-minimization algorithms and an application in robust face recognition: a review. In: Proceedings of IEEE International Conference on Image Processing, pp. 1849–1852 (2010)
10. Yang, A.Y., Sastry, S.S., Balasubramanian, A.G., Sastry, S.S., Ma, Y.: Fast 1-minimization algorithms for robust face recognition. IEEE Trans. Image Process. **22**(8), 3234–3246 (2013)
11. Beck, A., Teboulle, M.: A fast iterative shrinkage-thresholding algorithm for linear inverse problems. SIAM J. Image Sci. **2**(1), 183–202 (2009)
12. Gross R., Matthews I., Cohn J., Kanade T., Baker S.: Multi-PIE. In: Proceedings of IEEE International Conference on Automatic Face Gesture Recognition, pp. 1–8 (2008)
13. Nene S.A., Nayar S.K., Murase, H.: Columbia object image library (COIL-100). Technical Report CUCS-005-96 (1996)

Fashion Outfit Style Retrieval Based on Hashing Method

Yujuan Ding and Wai Keung Wong

Abstract This paper proposes an outfit retrieval method for moving fashion items dressed by models in the catwalk videos. The proposed method aims at retrieving similar style in fashion outfits images using a bilinear supervised hashing algorithm. The targeted images are labeled with the information of both color and attributes for comprehensive description of clothing style. The speed up robust features (SURFs) are extracted as the low-level features of fashion images and fed into the hashing algorithm as original data. To achieve better retrieval performance, a bilinear supervised hashing method is employed to learn high-quality hash codes. Outfits with similar style to the target are expected to be retrieved by hash code ranking. The experiment was conducted on a dataset composed of fashion outfit images and the experimental results show that the proposed method can retrieve styles which are close to query of color and attributes.

Keywords Fashion style retrieval · Fashion catwalk images · Hashing algorithm

1 Introduction

Recently, with great demand of fashion retail industry on enhancement of the customer shopping experience, extensive research has focused on clothing classification, attribute prediction, and clothing item retrieval [1]. With the development of computer science, increasing studies have been launched for clothing analysis based on the techniques of computer vision and pattern recognition [2]. Clothing recognition and retrieval has great research value since it can be used in industry in many aspects, for example, improve the online shopping recommendation [3]. In many research

Y. Ding (✉) · W. K. Wong
Institute of Textiles and Clothing, The Hong Kong Polytechnic University, Hong Kong, Hong Kong
e-mail: dingyujuan385@gmail.com

W. K. Wong
e-mail: calvin.wong@polyu.edu.hk

© Springer Nature Switzerland AG 2019
W. K. Wong (ed.), *Artificial Intelligence on Fashion and Textiles*,
Advances in Intelligent Systems and Computing 849,
https://doi.org/10.1007/978-3-319-99695-0_23

studies, the sources of fashion images for analysis are mainly from street photos, images in online store or social media platforms. However, very limited studies focus on catwalk images for analysis.

Fashion catwalk images deliver the design idea and show the newest fashion trend. They provide fashion designers and fashion buyers valuable resources for design inspiration and purchase plan. Nowadays, designers always review the catwalk videos or watch pictures to search information about previous, contemporary and most updated design for inspiration and mastering the fashion style trends and changes. However, such processes are daunting and time consuming, which makes it meaningful of making effort on fashion catwalk images analysis techniques. Developing the method for outfit retrieval in similar style not only improves fashion designers' work efficiency, but also gives ordinary people professional fashion style suggestion.

In this paper, a fashion outfit retrieval method focusing on a specific fashion dataset about catwalk images is proposed. The purpose is to find the most similar outfit in terms of style, which is seldomly studied in previous research. However, since the style is very abstract and hard to be defined or quantified, there is no ground truth of the similarity. This method is attempted to find the visually similar style for the given outfit in the fashion dataset. There are three main steps of the proposed method, namely fashion outfit description, low-level feature extraction and image retrieval. First, the color and attributes features are used to comprehensively describe fashion style. Then, to better represent the raw image of fashion outfit, low-level feature needs to be extracted, which is chosen as speed up robust feature (SURF) in this paper. Finally, for image retrieval algorithm, an effective and efficient bilinear supervised discrete hashing method is employed.

The rest of paper is organized as follows. Section 2 reviews related works about clothing description and retrieval. In Sect. 3, the fashion outfit retrieval method is introduced. The performance of the proposed method is validated by experiments in Sect. 4. Finally, some conclusion remarks are given in Sect. 5.

2   Related Works

Describing clothing appearance is the fundamental part of clothing related research with application of computer vision methods. Clothing is a rich visual domain in the standpoint of computer vision so that it is complicated to understand [4]. Therefore, many studies have been done to improve clothing visual understanding, especially detailed clothing attributes understanding. Chen et al. proposed a fully automated system to generate a list of common attributes of clothing [5]. There are in total 26 attributes considered in their research, including clothing pattern, major color, wearing necktie, collar presence and so on. They collected over 1000 pictures of dressed people to train their attributes classifier to make the system they proposed to describe the clothing appearance with semantic attributes. Di et al. conducted research specifically on female coat [6], and used 27 binary attributes that are specific to coat and jacket style to label images without considering all color features. They proposed retrieval method based on single attribute which only realized the attribute level retrieval. Chen et al. built a larger dataset with richer attributes and developed a system which can describe people based on clothing attributes [7].

Different from specific clothing attributes, the whole style of outfits is a relatively abstract concept and hard to quantify. Therefore, limited studies have been done on clothing style analysis. However, a few researchers still found the importance of this area and did some research works. Jia jia et al. built a fashion semantic space based on the image-scale space in aesthetic area to promote the understanding of clothing fashion style [8]. Before this study, their group had done some work on aesthetic features about clothing [9]. Kiapour et al. built a dataset related to style recognition named Hipster War, in which they manually defined and labeled the style of each image in the dataset. They tried to associate the elements of clothing with each style and conducted a style rating game [10]. Based on the same dataset, Edgar Simo-Serra et al. employed deep learning techniques to extract image features, and the performance of style retrieval on the dataset was improved by doing so [11].

3 Methodology

The problem of style retrieval of outfit contains three key steps, which are outfit description, low-level features extraction and style retrieval algorithm.

3.1 Joint Description for Clothing with Color and Attributes

The visual similarity of images is based on two aspects: color and shape [12] which have great influence on the style and people's intuitive emotional feeling on the outfits. In this study, both color and shape attributes will be used to label images for a more comprehensive description of outfits' style.

In existing research, the color feature is usually ignored to simplify the clothing description model. However, color is an important part of fashion outfit design as it influences the first impression of human to a clothing item. The five-color theme is the well-known and common used strategy to describe the general color pattern of images which consists of five dominant colors of image. In this paper, since the research objects are outfits, the five-color theme of outfit is adopted as the color descriptor [13], which can weaken the effect of background. Some examples of five-color theme of fashion outfit images are shown in Fig. 1.

Besides color, clothing attributes including collar shape, sleeve length and so on, is another key factor that affects the style. A large amount of research mentioned in Sect. 2 also proves the importance of attributes for describing clothing. Since many studies have been done in this area, the attributes listed in [14] are adopted with some minor modification is adopted in this study. Specifically, the whole outfit is divided into top and bottom and attributes of which are separately defined. The total

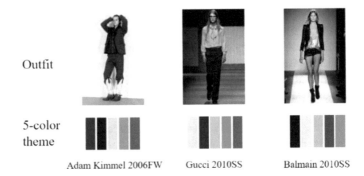

Fig. 1 Examples of five-color theme of outfits

categories of attributes considered in this paper is 21, nine of them are top attributes and 12 of them are bottom attributes.

3.2 Hashing Algorithm for Style Retrieval of Outfit

To conduct the fashion style retrieval, the low-level feature needs to be extracted first for better representation of raw images. The feature should be with lower dimensionality but can reserve as much information in the raw image. Considering the effectiveness of the feature and the computational cost, SURF is adopted as the low-level feature for the fashion catwalk images in this research for further analysis.

For image retrieval, hashing is a typical method which learns binary codes from low-level features of images. The learned codes need to keep enough original information of images for retrieval. In the past few years, various hashing methods were proposed in literature [15–19], they can not only save the computing cost, but also achieve good performance in retrieval precision. In this paper, considering the low-level feature adopted and to take advantage of the label information of the image, a bilinear supervised hashing method is adopted to undertake the retrieval task. The objective function of the hashing method is as follows:

$$
\min_{B,W,Q_1,Q_2} \left[\|Y - W^T B\| + \lambda\|W\| + \upsilon \sum_{i=1}^{n} \|D_i - Q_1^T X_i Q_2\| + \alpha\|Q_1\| + \beta\|Q_2\| \right],
$$
$$
\text{s.t. } B = [b_1, b_2, \ldots b_n] \in \{-1, 1\}^{c \times n} \; b_i = \text{vec}(D_i) \text{ for } i = 1 : n \tag{1}
$$

where X is the original data, Y is the label matrix and B is the binary matrix which needs to learn. Q_1, Q_2 are two mapping matrices for X and W is the mapping matrix for B. λ, υ, α, and β are four parameters to adjust the importance of every term. The special matrix D in the objective function is the binary matrix for each set of data. For the optimization problem, specific discrete method is used to solve it. Detailed process will not be stated in this paper due to the length limitation, they can be found in author's other papers. The results of the optimization are given, W, Q_1 and Q_2 can be obtained according to following equation.

$$W = (BB^T + \lambda I)^{-1} BY^T$$

$$Q_1 = \left(\upsilon \sum_{i=1}^{n} X_i Q_2 Q_2^T X_i^T + \alpha I \right)^{-1} \left(\upsilon \sum_{i=1}^{n} X_i Q_2 D_i^T \right) \tag{2}$$

$$Q_2 = \left(\upsilon \sum_{i=1}^{n} X_i^T Q_1 Q_1^T X_i + \beta I \right)^{-1} \left(\upsilon \sum_{i=1}^{n} X_i^T Q_1 D_i \right) \tag{3}$$

B is different with other three variables because of the binary constraint, the strategy used here to update B is bit-by-bit updating. Let z be a column of B and z is iteratively updated by the following equation:

$$z = \text{sgn}\left(m^T - \upsilon^T W'^T B' \right), \tag{4}$$

where m and v are corresponding columns of M and W' $M = Y^T W^T + \upsilon Q^T$. W', B' are rest matrix of W and B.

The proposed method aims at conducting hash coding for every image and evaluating the similarity between different images by calculating the distance of hash codes. To achieve this objective, the hash codes are required to keep information of outfit in images and this information should enough to represent raw images. In order to learn high-quality binary codes, a certain number of well-labeled images are needed to train the binary learning algorithm. After that, the trained model can be employed to code new images. For an outfit image query, the algorithm is expected to retrieve the most similar outfits in style by comparing their hash codes. The overall process of the proposed retrieval method is shown in Fig. 2.

4 Experiment

4.1 Dataset

The fashion outfit dataset built by Ma et al. [14] is adopted in this paper to validate the performance of the proposed fashion style retrieval method. This dataset consists of 32,133 full-body fashion show images including men's and women's. All images are well-labeled with two outfit features, namely color and attributes. Specifically, five-color theme of clothing is extracted as the color feature and 21 attributes including clothing length, collar shape, etc., are chosen as the key attributes. Color and attributes

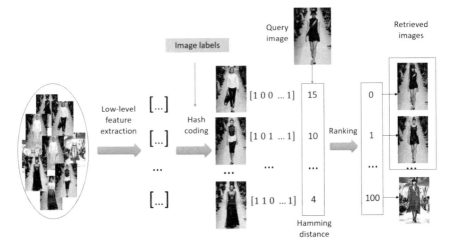

Fig. 2 The philosophy of hashing-based fashion style retrieval

form the label vector of images. The dimension of the label vector for each picture is 161. More detailed information of attributes used in this dataset can be found in [14]. We used majority of images in the dataset for training and randomly left a hundred images as queries to test the performance of the proposed method.

4.2 Performance of Fashion Style Retrieval Method

Based on the procedures mentioned in Sect. 4.3, the first step is to extract low-level image features, which are SURFs in this paper. When the low-level features are well prepared, they will be fed as the original data to train the hash function. To find the better parameter setting in the objective function, we tried 90 combinations and compared their performance. Specifically, we conducted a retrieval experiment for every parameter setting and recorded their average retrieval precision on the dataset CIFAR 10. The combination with best results was selected as the parameter setting in the outfit retrieval experiment. $\lambda, \upsilon, \alpha, \beta$ are finally set to $10^{-3}, 10^{-7}, 10^{-2}, 10^{-2}$ respectively. Besides, the size of hash codes, namely D_i in the objective is set to 8×4.

As mentioned in Sect. 3, the retrieval algorithm is capable to find similar outfits of the query by calculating the hamming distance of their hash codes. Exemplary results of retrieved similar style outfits and the reference one are shown in Fig. 3.

Query   Retrieved similar style

Fig. 3 Examples of fashion style retrieval

4.3 Analysis

From Fig. 3, it can be seen that the results for seeking similar outfits are satisfactory on the whole style. This proves the effectiveness of our retrieval method. Since both attributes and colors are considered for describing images, the learned binary codes contain both information, which is obvious shown in Fig. 3. Besides, comparing with results of the six samples, it can be seen that the attribute and color make unequal contribution in binary codes learning in different results. In other words, the algorithm itself is trying to balance the impact of these two to obtain more representative codes for outfits. In summary, the proposed method can identify similar style based on color and attributes.

5 Conclusion

This paper proposed an outfit retrieval method for retrieving similar style based on a bilinear supervised hashing algorithm. The method focuses on specific research objects which are fashion catwalk images. The targeted images are labeled with the information of five-color theme and clothing attributes for comprehensive description of style. In the retrieval process, the low-level feature SURFs are extracted and then fed into the hashing algorithm as original data. The hashing algorithm is adopted to learn high-quality binary codes which can reflect the style of outfits in images. The retrieval task is finally realized by hash code ranking.

Experiment was conducted on a dataset composed entirely of fashion outfit images. Experimental results show that the proposed method can find out styles which are very similar with query in the respect of color and attributes.

Acknowledgements This paper was supported by the Hong Kong Polytechnic University.

References

1. Liu, Z., Luo, P., Qiu, S., Wang, X., Tang, X.: Deepfashion: powering robust clothes recognition and retrieval with rich annotations. In: Proceedings of the IEEE Conference on Computer Vision and Pattern Recognition, pp. 1096–1104 (2016)
2. Yang, W., Luo, P., Lin, L.: Clothing co-parsing by joint image segmentation and labeling. In: Proceedings of the IEEE Conference on Computer Vision and Pattern Recognition, pp. 3182–3189 (2014)
3. Liu, S., Song, Z., Liu, G., Xu, C., Lu, H., Yan, S.: Street-to-shop: cross-scenario clothing retrieval via parts alignment and auxiliary set. In 2012 IEEE Conference on Computer Vision and Pattern Recognition, pp. 3330–3337 (2012)
4. Matzen, K., Bala, K., Snavely, N.: StreetStyle: exploring world-wide clothing styles from million of photos (2017)
5. Chen, H., Gallagher, A., Girod, B.: Describing clothing by semantic attributes. In: European Conference on Computer Vision, pp. 609–623 (2012)
6. Di, W., Wah, C., Bhardwaj, A., Piramuthu, R., Sundaresan, N.: Style finder: fine-grained clothing style detection and retrieval. In: 2013 IEEE Conference on, Computer Vision and Pattern Recognition Workshops (CVPRW), pp. 8–13 (2013)
7. Chen, Q., Huang, J., Feris, R., Brown, L.M., Dong, J., Yan, S.: Deep domain adaptation for describing people based on fine-grained clothing attributes. In: 2015 IEEE Conference on Computer Vision and Pattern Recognition (CVPR), pp. 5315–5324 (2015)
8. Jie, X., Mok, P.Y., Yuen, C.W.M., Yee, R.W.Y.: A web-based design support system for fashion technical sketches. Int. J. Clothing Sci. Technol. **28**, 130–160 (2016)
9. Jia, J., Huang, J., Shen, G., He, T., Liu, Z., Luan, H.-B. et al.: Learning to appreciate the aesthetic effects of clothing. In: AAAI, pp. 1216–1222 (2016)
10. Kiapour, M.H., Yamaguchi, K., Berg, A.C., Berg, T.L.: Hipster wars: discovering elements of fashion styles. In: European Conference on Computer Vision, pp. 472–488 (2014)
11. Simo-Serra, E., Ishikawa, H.: Fashion style in 128 floats: joint ranking and classification using weak data for feature extraction. In: 2016 IEEE Conference on Computer Vision and Pattern Recognition (CVPR), pp. 298–307 (2016)
12. Wang, X., Jia, J., Yin, J., Cai, L.: Interpretable aesthetic features for affective image classification. In: 2013 IEEE International Conference on Image Processing, pp. 3230–3234 (2013)

13. Wang, B., Yu, Y., Wong, T.-T., Chen, C., Xu, Y.-Q.: Data-driven image color theme enhancement. ACM Trans. Graph. **29**, 1–10 (2010)
14. Ma, Y., Jia, J., Zhou, S., Fu, J., Liu, Y., Tong, Z.: Towards better understanding the clothing fashion styles: a multimodal deep learning approach. In: AAAI, pp. 38–44 (2017)
15. Cui, Y., Jiang, J., Lai, Z., Hu, Z., Wong, W.: Supervised discrete discriminant hashing for image retrieval. Pattern Recognit. **78**, 79–90 (2018)
16. Shen, F., Shen, C., Liu, W., Tao Shen, H.: Supervised discrete hashing. In: Proceedings of the IEEE Conference on Computer Vision and Pattern Recognition, pp. 37–45 (2015)
17. Wang, J., Zhang, T., Sebe, N., Shen, H.T.: A survey on learning to hash. IEEE Trans. Pattern Anal. Mach. Intell. **40**, 769–790 (2018)
18. Weiss, Y., Torralba, A., Fergus, R.: Spectral hashing. In: Advances in Neural Information Processing Systems, pp. 1753–1760 (2009)
19. Gong, Y., Kumar, S., Rowley, H.A., Lazebnik, S.: Learning binary codes for high-dimensional data using bilinear projections. In: 2013 IEEE Conference on Computer Vision and Pattern Recognition (CVPR), pp. 484–491 (2013)

Supervised Locality Preserving Hashing

Xiao Zhou, Zhihui Lai and Yudong Chen

Abstract Hashing methods are becoming increasingly popular because they can achieve fast retrieval of large-scale data by representing the images with binary codes. However, the traditional hashing methods tend to obtain the binary codes by relaxing the discrete problems which greatly increase the information loss. In this paper, we propose a novel hash learning method, called Supervised Locality Preserving Hashing (SLPH) for image retrieval. Different from the traditional two-steps methods which learn low-dimensional features and binary codes of the data separately, we directly obtain the binary codes and thus reduce the information loss. Besides, we add graph-regularized learning on the designed model to avoid over-fitting and improve the performance. Experiments on two benchmark databases show that the proposed SLPH performs better than some state-of-the-art methods.

Keywords Linear regression · Binary code · Feature extraction · Hash learning

1 Introduction

Hashing methods include two main categories: data-independent methods and data-dependent methods.

Data-independent hashing methods use random projections to map the original data into low-dimensional binary spaces. As a representative data-independent method, Locality-Sensitive Hashing (LSH) [1] has been well improved by utilizing different distance measures such as, p-norm distance [2], cosine similarity [3], Mahalanobis metric [4] and kernel similarity [5, 6]. However, LSH-based methods require long binary codes to achieve well performance.

In contrast, data-dependent or learning-based hashing methods try to learn more sophisticated binary codes by utilizing training data. Generally, data-dependent

X. Zhou · Z. Lai (✉) · Y. Chen
The College of Computer Science and Software Engineering,
Shenzhen University, Shenzhen 518060, China
e-mail: lai_zhi_hui@163.com

© Springer Nature Switzerland AG 2019
W. K. Wong (ed.), *Artificial Intelligence on Fashion and Textiles*,
Advances in Intelligent Systems and Computing 849,
https://doi.org/10.1007/978-3-319-99695-0_24

hashing methods can be divided into two directions, i.e., linear hashing methods and nonlinear hashing methods. The former methods attempt to find a set of hyperplanes as linear hash functions. The representative methods include PCA Hashing [7], Iterative Quantization (ITQ) [8], Isotropic Hashing [9], Supervised Minimal Loss Hashing (MLH) [10], Semi-Supervised Hashing (SSH) [7], LDA Hashing [11], Ranking-Based Supervised Hashing [12] and FastHash [13], etc. Besides, some nonlinear methods introduce kernel function to improve the performance, such as Binary Reconstructive Embedding (BRE) [14], Random Maximum Margin Hashing (RMMH) [15], Kernel-Based Supervised Hashing (KSH) [16], and the kernel version of ITQ [8].

However, these methods fail to discover the manifold structure of datasets. Some manifold-based methods, such as Spectral Hashing (SH) [17], Anchor Graph Hashing (AGH) [18, 19] and Inductive Manifold Hashing (IMH) [20, 21] embed the binary codes to low-dimensional manifold space. Therefore, the latent structure in the original space can be preserved. As indicated in [22], manifold structure of data plays an important role in feature selection tasks. However, many existing methods only consider the global structure of the data and the performance is hard to guarantee.

Therefore, to integrate the locality preserving property and the hash learning to obtain good hash code for effective linear classification, in this paper, we propose a novel Supervised Locality Preserving Hashing method for hash code learning (SLPH). SLPH jointly learns discriminative information and local structure of the data. The main contributions of this paper can be summarized as follows:

(1) Different from [23], the proposed SLPH takes the local structure of the data into consideration to preserve the neighborhood relationship of the data in low-dimensional space, and it can improve the performance of many tasks in discriminant subspace learning, manifold learning and hash learning. To directly obtain the optimal solution, a discrete optimization algorithm is also proposed.

(2) Comprehensive evaluations are conducted on two large-scale datasets. It is shown that the proposed SLPH performs better than the state-of-the-art methods.

The rest of the paper is organized as follows. In Sect. 2, the SLPH is proposed. In Sect. 3, we evaluate our method on two large-scale datasets. The conclusions are given in Sect. 4.

2   Supervised Locality Preserving Hashing

In this section, we will introduce the proposed Supervised Locality Preserving Hashing (SLPH) method.

2.1 The Objective Function of SLPH and Its Formulation

Given n instances $X = \{x_i\}_{i=1}^n$, we aim to learn a set of hash code $B = \{b_i\}_{i=1}^n \in \{-1, 1\}^{l \times n}$ to preserve their similarities in the binary space, where vector b_i is the l-bits hash codes of x_i. s is a similarity matrix which characterizes the likelihood of data points. If data pair (x_i, x_j) is similarity, $S_{ij} = 1$ and 0 otherwise.

The objective function of SLPH is as follows:

$$\min_{B, W, P} \sum_{i=1}^n \sum_{j=1}^n \left\| b_i - W^T x_j \right\|_2^2 S_{ij} + \beta \sum_{i=1}^n \left\| b_i - P x_i \right\|_2^2 + \alpha \|W\|_F^2 + \gamma tr(W^T X S X^T W)$$

$$\text{s.t. } b_i \in \{-1, 1\}^L.$$

$$(1)$$

where $\|\cdot\|_F$ is the Frobenius norm of a matrix, p and W are projection matrix. The idea of designing model (1) is as follows: First, we hope to preserve the similarity within classes in the binary spaces by using the first term. Then, we add the second term to learn a projection matrix for hash learning so as to reduce the information loss. Finally, we use two regularization term including the traditional L_2-norm regularized learning and locality preserving graph-regularized learning to avoid over-fitting. The optimization algorithm is displayed in next subsection.

2.2 The Solutions of SLPH

The problem formulated in (1) is a combined optimization problem with three variables. We use an alternatively iterative algorithm to optimize them respectively.

P-step: If B and w are fixed, we have following minimization problem

$$\min_{P} \|B - PX\|_F^2$$

$$\text{s.t. } B \in \{-1, 1\}^{L \times n}.$$

$$(2)$$

Taking partial deviation with respect to P to be zero, we can derive:

$$P = (X^T X)^{-1} X^T B$$

$$(3)$$

W-step: When B and P are fixed, we need to solve

$$\min_{W} n \sum_{i=1}^n \sum_{j=1}^n \left\| b_i - W^T x_j \right\|_2^2 S_{ij} + \alpha \|W\|_F^2 + \gamma tr(W^T X S X^T W)$$

$$\text{s.t. } b_i \in \{-1, 1\}^L.$$

$$(4)$$

Taking partial deviation with respect to w to be zero, we further have:

$$W = (DXX^T + \alpha I + \gamma XSX^T)^{-1} BXS, \tag{5}$$

where D is a diagonal matrix with its entries being the column or row (S is symmetric) sums of S, i.e., $D_{ii} = \sum_{j=1}^{n} S_{ij}$.

B-step: With all variables but B fixed, the solution can be update by solving the following optimization problem:

$$\min_{B} \sum_{i=1}^{n} \sum_{j=1}^{n} \left\| b_i - W^T x_j \right\|_2^2 S_{ij} + \beta \sum_{i=1}^{n} \left\| b_i - P x_i \right\|_2^2 \tag{6}$$

$$\text{s.t. } b_i \in \{-1, 1\}^L.$$

which has a closed-form solution:

$$B = \text{sgn}(W^T XS + \beta P^T X) \tag{7}$$

B is updated by using a single step.

3 Experiments

In this section, we investigate SLPH's performance by conducting experiments on a computer with an Intel Xeon processor (2.20 GHz), 128-GB RAM, and configured with Microsoft Windows 7 and MATLAB 2017a.

A set of experiments are presented to show the effectiveness of the proposed hashing method against several state-of-the-art supervised hashing methods including SDH and FSDH [24] and unsupervised methods including LSH, PCA-ITQ, and SH. There are two large-scale datasets used for evaluating the above methods, Fashion-MNIST and CIFAR-10. Besides, hamming ranking (mean of average precision, MAP), hash lookup (precision, recall, and F-measure of Hamming radius 2) and accuracy are employed for the evaluation of above methods.

We use the public codes and the parameters suggested by the corresponding authors. For SLPH, we empirically set α to be 1, β to be 1 and γ to 1e-5; the maximum iteration number t is set to be 20. Similar to SDH and FSDH, we preprocessed the samples by using kernel technique and 1000 anchor points were utilized.

3.1 Data Sets Details

CIFAR-10 dataset contains 60,000 32×32 color images, with 6000 images for each class. Each image is represented by a 512-dimensional GIST feature vector.

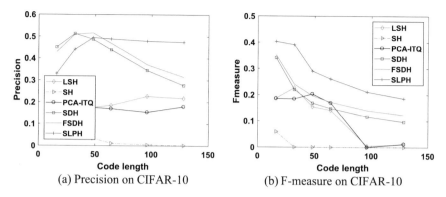

(a) Precision on CIFAR-10   (b) F-measure on CIFAR-10

Fig. 1 Precision and F-measure versus the number of binary bits (16, 32, 48, 64, 96, 128) on CIFAR-10 database

Fashion-MNIST dataset consists of 28×28 grayscale images of 70,000 fashion products from 10 categories, and there are 7000 in each category. It is split into a training set of 60,000 samples and a test set with 1000 examples.

3.2 Experimental Results Comparisons

Results on CIFAR-10. We compare the hash lookup results in Precision and F-measure which can be found in Fig. 1. Compared with other methods, SLPH performs better in longer codes as shown in Fig. 1a. When the code length is 128, the best Precision performance of proposed hash method is 47.43%, compare with 21.72% for LSH, 17.97% for PCA-ITQ, 0.1% for SH, 27.64% for SDH and 31.46% for FSDH. In Fig. 1b, SLPH has a remarkable improvement in terms of F-measure. Though most of the methods decrease dramatically with the increase of code length, SLPH still has advantages over them. When the code length is 16, SLPH performs the best. Thus, the proposed SLPH outperforms other methods in Precision and F-measure.

Results on Fashion-MNIST. The experiment of Precision is shown in Table 1. Although SDH performs best when the number hashing bits equals 48, we can see that SLPH achieves the best of three effects in total. Furthermore, as the code length increases, the advantage of SLPH is more obvious.

Table 1 Comparison of the precision of different code length on fashion-MNIST

Method	48	64	96	128
SLPH	0.8012	**0.8021**	**0.7936**	**0.7842**
SDH	**0.8027**	0.8011	0.7835	0.7737
FSDH	0.7992	0.7960	0.7834	0.7721
SH	0.6959	0.5842	0.2771	0.1030
PCA-ITQ	0.5934	0.6541	0.6825	0.0065
LSH	0.5621	0.6476	0.6321	0.6253

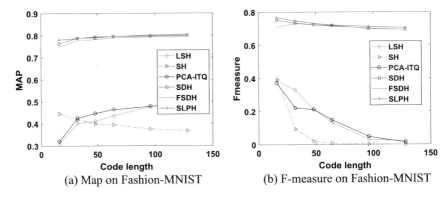

(a) Map on Fashion-MNIST   (b) F-measure on Fashion-MNIST

Fig. 2 Map and F-measure versus the number of binary bits (16, 32, 48, 64, 96, 128) on Fashion-MNIST database

From Fig. 2a, it shows the comparison results on Fashion-MNIST dataset in Recall (for hash lookup). We can see that SLPH achieves the best performance at almost all the code lengths, except the shortest one. To be specific, SLPH with large codes has a remarkable improvement. However, when the code length becomes large, SLPH is very similar to SDH and FSDH in F-measure, while obviously outperforms LSH, SH and CCA-ITQ. In Fig. 2b, the best F- measure rate of the proposed SLPH is 76.64%, when the number of hashing bits is 16, compared with 38.58% for LSH, 39.35% for SH, 37.02% for PCA-ITQ, 75.18% for SDH and 70.92% for FSDH. Thus, SLPH best performs than all other hashing methods in this case.

3.3 Convergence of the Algorithm

Since we design and iterative algorithm to reduce the objective function values, it is necessary to test the convergence of the algorithm. Figure 3 displays the convergence of our algorithm on CIFAR-10 and Fashion-MNIST databases. We can find that the algorithm converges very fast on both databases.

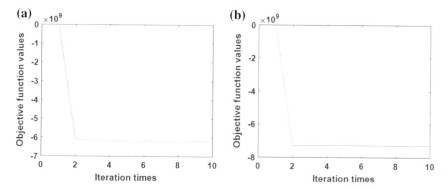

Fig. 3 Convergence on **a** CIFAR-10 and **b** fashion-MNIST databases

4 Conclusion

In this paper, we propose a supervised discrete model to directly learn the binary codes. We design an iterative algorithm by decomposing the optimization problem into three sub-problems. Extensive experimental results on two large-scale ben datasets demonstrate the performance of the proposed method.

Acknowledgements This work was supported in part by the Natural Science Foundation of China (Grant 61573248, Grant 61773328, Grant 61732011 and Grant 61703283), Research Grant of The Hong Kong Polytechnic University (Project Code: G-UA2B), China Postdoctoral Science Foundation (Project 2016M590812 and Project 2017T100645), the Guangdong Natural Science Foundation (Project 2017A030313367 and Project 2017A030310067), and Shenzhen Municipal Science and Technology Innovation Council (No. JCYJ20170302153434048 and No. JCYJ20160429182058044).

References

1. Gionis A., Indyk P., Motwani R.: Similarity search in high dimensions via hashing. In: Proceedings of the 25th International Conference on Very Large Data Bases, 1999, pp. 518–529 (1999)
2. Datar M., Immorlica N., Indyk P., Mirrokni V.S.: Locality-sensitive hashing scheme based on p-stable distributions. In: Twentieth Symposium on Computational Geometry, vol. 34, No. 2, pp. 253–262 (2004)
3. Charikar M.S.: Similarity estimation techniques from rounding algorithms. In: Proceedings of the 34th STOC, 2002, pp. 380–388 (2002)
4. Kulis, B., Jain, P., Grauman, K.: Fast similarity search for learned metrics. IEEE Trans. Pattern Anal. Mach. Intell. **31**, 2143–2157 (2009)
5. Kulis B., Grauman K.: Kernelized locality-sensitive hashing for scalable image search. In: IEEE International Conference on Computer Vision, 2010, pp. 2130–2137 (2010)
6. Raginsky M.: Locality-sensitive binary codes from shift-invariant kernels. In: Advances in Neural Information Processing Systems, pp. 1509–1517 (2009)

7. Wang, J., Kumar, S., Chang, S.F.: Semi-supervised hashing for large-scale search. IEEE Trans. Pattern Anal. Mach. Intell. **34**, 2393–2406 (2012)
8. Gong, Y., Lazebnik, S., Gordo, A., Perronnin, F.: Iterative quantization: a procrustean approach to learning binary bodes for large-scale image retrieval. IEEE Trans. Pattern Anal. Mach. Intell. **35**, 2916–2929 (2013)
9. Kong W., Li W.J.: Isotropic hashing. In: International Conference on Neural Information Processing Systems, 2012, pp. 1646–1654 (2012)
10. Norouzi M., Fleet D.J.: Minimal loss hashing for compact binary codes. In: International Conference on Machine Learning, 2011, pp. 353–360 (2011)
11. Strecha, C., Bronstein, A., Bronstein, M., Fua, P.: LDAHash: improved matching with smaller descriptors. IEEE Trans. Pattern Anal. Mach. Intell. **34**, 66–78 (2012)
12. Wang J., Liu W., Sun A.X., Jiang Y.G.: Learning hash codes with listwise supervision. In: IEEE International Conference on Computer Vision, 2014, pp. 3032–3039 (2014)
13. Lin G., Shen C., Shi Q., Hengel A., Suter D.: Fast supervised hashing with decision trees for high-dimensional data. In: Proceedings of the IEEE Conference on Computer Vision and Pattern Recognition, 2014, pp. 1971–1978 (2014)
14. Kulis B., Darrell T.: Learning to hash with binary reconstructive embeddings. In: Proceedings of the 22nd International Conference on Neural Information Processing Systems, 2009, pp. 1042–1050 (2009)
15. Buisson O., Buisson, O.: Random maximum margin hashing. In: Proceedings of the IEEE Conference on Computer Vision and Pattern Recognition, 2011, pp. 873–880 (2011)
16. Chang S.F.: Supervised hashing with kernels. Supervised hashing with kernels. In: Proceedings of the IEEE Conference on Computer Vision and Pattern Recognition, 2012, pp. 2074–2081 (2012)
17. Weiss Y., Torralba A., Fergus R.: Spectral hashing. In: Proceedings of the 21st International Conference on Neural Information Processing Systems, 2008, pp. 1753–1760 (2008)
18. Belkin M., Niyogi P.: Laplacian Eigenmaps for dimensionality reduction and data representation. MIT Press (2003)
19. Liu W., Kumar S., Kumar S., Chang S.F.: Discrete graph hashing. In: Proceedings of the 27th International Conference on Neural Information Processing Systems, 2014, pp. 3419–3427 (2014)
20. Shen F., Shen C., Shi Q., Hengel A., Tang Z.: Inductive hashing on manifolds. In: Proceedings of the IEEE Conference on Computer Vision and Pattern Recognition, 2013, pp. 1562–1569 (2013)
21. Shen, F., Shen, C., Shi, Q., Hengel, A., Tang, Z., Shen, H.T.: Hashing on nonlinear manifolds. IEEE Trans. Image Process. **24**, 1839–1851 (2015)
22. He, X., Niyogi, P.: Locality preserving projections. Adv. Neural. Inf. Process. Syst. **16**, 186–197 (2003)
23. Shen F., Shen C., Liu W., Shen H.T.: Supervised discrete hashing. In: Proceedings of the IEEE Conference on Computer Vision and Pattern Recognition, 2015, pp. 37–45 (2015)
24. Gui, J., Liu, T., Sun, Z., Tao, D., Tan, T.: Fast supervised discrete hashing. IEEE Trans. Pattern Anal. Mach. Intell. **40**(2), 490–496 (2018)

Co-designing Interactive Textile for Multisensory Environments

H. Y. Kim, J. Tan and A. Toomey

Abstract In Hong Kong, the number of people aged 65 years and above is rapidly increasing and the population of those in this age range is projected at 2.3 million in 2036 (Yu et al. in Dementia trends: impact of the ageing population and societal implications for Hong Kong. Hong Kong Jockey Club, Hong Kong, 2010 [1]). Given the rapid increase in the population of people with dementia, the design of a dementia-friendly environment in care facilities has become crucial for the enhanced quality of life of these people. A multisensory environment (MSE) is a term used to describe a multisensory space that can be used to provide sensory stimulation or reduce sensory demand, thereby increasing engagement and reducing challenging behaviour (Lesley and Anke in HERD 10(5): 39–51, 2016 [2]). Currently, MSE tools are relevant to people with late-stage dementia, whereas people in the early stages of this condition have limited association with such tools. The existing tools are generic and emphasises simple functions instead of design. This study reviews the existing design processes for MSEs and analyses the preliminary data derived from a pilot design study involving the Hong Kong Sheng Kong Hui Lok Man Alice Kwok Integrated Service Centre.

Keywords Interactive textiles · People with dementia · Co-design Multisensory environment

1 Design Process in a Multisensory Environment

Multisensory environments (MSEs) aim to increase engagement and reduce behaviours perceived as challenging, thereby providing sensory stimulation and reducing sensory demand [2]. The development of MSEs has emerged from an identi-

H. Y. Kim (✉) · J. Tan
Hong Kong Polytechnic University, Hong Kong, Hong Kong
e-mail: heeyoung.kim@connect.polyu.hk

A. Toomey
Royal College of Art, London, UK

© Springer Nature Switzerland AG 2019
W. K. Wong (ed.), *Artificial Intelligence on Fashion and Textiles,*
Advances in Intelligent Systems and Computing 849,
https://doi.org/10.1007/978-3-319-99695-0_25

fied need to modify the environment for those who are severely disabled, particularly to stimulate the primary senses, thereby generating pleasurable sensory experiences in an atmosphere of trust and relaxation without the need for intellectual activities [3]. The use of multisensory rooms for the elderly with dementia is becoming an increasingly popular adjunct to caregiving [4]. Multisensory rooms facilitate the reduction of agitation and anxiety, although such facility can also engage and delight users, stimulate reactions and encourage communication [5]. However, limited studies on MSEs have been conducted, particularly those that considers the actual design, including functionality and aesthetics, and its impact on engagement and well-being [6]. Accordingly, the nature of the complex and diverse symptoms of dementia has necessitated the personalisation of sensory tools for individuals based on their sensory preferences. This study reviews the designs of researchers who have explored MSEs from the designer's perspective. The reason for such analysis is that only a few studies currently discuss MSEs from the aforementioned perspective.

1.1 Katie Gaudion: Affordable Sensory Props in MSEs for Adults with Autism

Katie Gaudion is a senior design researcher in the Helen Hamlyn Design Centre of the Royal College of Art and has a background in textile design. Her research focuses on the challenge to build MSEs using inexpensive materials and create sensory textile props using inexpensive daily objects and materials through a participatory design process for adults with autism. The aim of her research is to stimulate the senses of these adults, encourage them to be involved in the design process and understand their sensory preferences and needs through empathy. Gaudion critically explained that the majority of the design projects that involve people with autism in the design process intend to develop a new technology. Hence, researchers should be aware of the differences between the physical and virtual environments, the different challenges that they present to people with autism and the designer and how such challenges can influence the design process [7]. However, technology can generate multiple solutions for the problems related to the shortage of manpower and space to operate MSEs and maintenance of the primary role of such environment.

1.2 Sari Hedman: Applying Preferred Experiences in MSEs for People with Dementia

Sari Hedman is a Finnish artist and care practitioner who insists that MSEs should have supplementary design elements for aesthetic pleasure and safety and to revitalise the accessibility of current MSE tools. Currently, the traditional MSE tools are similar to children's toys that are age-inappropriate. Hedman explained that horse hair and brushes hang on walls for tactile stimulation but had been loosened from the context, thereby leading her to insist that multisensory rooms should activate reminiscence, relax, and cherish senses [8]. Hedman investigated the preferred experiences of people with dementia and applied the data of their preferences to build the sensory environments in six dementia care home settings in Helsinki.

1.3 Anke Jakob: Guidelines to Improve Sensory Environment for People with Dementia

Anke Jakob is a design research fellow in Kingston University and has a background in textile design. Her research involves designing therapeutic and sensory-experienced environment for people with dementia using such elements as light, materials and digital technology. Jakob's research aims to improve the sensory environment by providing advice to informal caregivers and professionals on the method of building a sensory environment in the home care setting to benefit those with dementia. Moreover, her collaborative research project with Collier [12] investigated the process of increasing the benefits of MSEs through the design of dementia care home settings. Jakob suggested that the design principles for MSEs should include feelings of comfort and safety, meaning and familiarity, multisensory experience, stimulation and relaxation, control and interaction, age appropriate and usable [9]. However, the majority of the research on multisensory tools are for late-stage dementia, whereas those involving early-stage dementia are limited.

A review of the design research on MSEs indicates that the design criteria on building a sensory room with existing materials and sensory tools is focused on the design development of sensory tools, particularly for people with dementia. Hence, the accessibility and usability of sensory tools should also be considered for the design development based on the needs of people with dementia.

2 MSE for People with Dementia

Dementia is a condition that comprises several symptoms, including the reduced ability to perform familiar tasks; impairment of memory, judgement and reasoning and changes in mood and behaviour [10]. Dementia is a general term used to describe the symptoms when the brain is affected by several conditions, most commonly Alzheimer's [11]. The current research focuses on Alzheimer's because it is the most common disease among people with dementia, particularly the mild to moderate-stage ones.

In Hong Kong, the number of people aged 65 years and above is rapidly increasing and the population of this age range is projected at 2.3 million in 2036 [1]. Accordingly, the rapid increase in the population of people with dementia necessitates the design of dementia-friendly environments, particularly dementia care facilities, to enhance their quality of life. The use of MSEs for people with dementia as a resource for meaningful engagement has beneficial effects, including the enhancement of job satisfaction and well-being of care givers and staff members and improvement in their interpersonal relationship with people with dementia [9]. Currently, dementia care facilities in Hong Kong are also equipped with MSE tools. However, these facilities appear to fail in addressing the specific needs of people with dementia because of inadequate design and poor facilitation, thereby resulting in the staff members becoming discouraged because of the perception that the space has limited value and eventually leads to the facility becoming unused [9].

2.1 MSE in Hong Kong

Collier reported that MSEs can encourage people with dementia to experience the therapeutic effects of sensory engagement activities [12]. The multifaceted tactile quality of textiles, such as the combination of textiles and lighting, can provide haptic and bodily experiences [13]. Selected free-standing, hanging or wall-mounted lights (e.g. fibre optic curtain, bubble wall, LED net) within easy view of users and hand-held equipment (e.g. fibre optic strands and glowing balls) create visual focus points [9]. The Hong Kong Sheng Kong Hui (HKSKH) Li Ka Shing Care & Attention Home for the Elderly in Shum Shui Po in Hong Kong also has a multisensory room with illuminative equipment, such as fibre optics, bubble tubes and starry lights (see Fig. 1).

The fibre optic tool in MSE enables people with dementia to touch the lighting, thereby breaking the perceptual barrier. This tool is sufficient to stimulate the sense of vision, although it remains limited in terms of tactile quality, which will provide soft and affirmative haptic experience. Moreover, the majority of the illuminative tools are equipped with a single function that demands additional staff resources to access. However, accessibility to MSE in a dementia care facility in Hong Kong is limited because of lack of space and manpower. The form of interactive textile, which is a fusion of soft materials and embedded electronics, can enhance the sensory experience of people with dementia. Sensory textiles can encourage informal caregivers and family members to engage in meaningful communication with people with dementia [14]. Modern technology can be considerably beneficial in creating age-appropriate switches to control sensory environments [9]. The addition of embedded electronics has enabled further personalization, while simple intuitive haptic electronic interfaces also facilitate the extension of sensory functions [15].

Fig. 1 Illuminative
equipment in the
multi-sensory room in
HKSKH Li Ka Shing Elderly
Care and Attention Home in
Shum Shui Po, Hong Kong.
Source Kim [16]

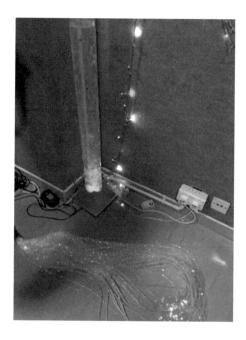

3 Preliminary Study Using Interactive POF Textiles

This preliminary study co-designs interactive POF textiles with staff members and occupational therapists for people with dementia in the HKSKH Lok Man Alice Kwok Integrated Service Centre. Co-design is a valuable method of addressing the complex needs of people with dementia to develop the appropriate designs to support their well-being [14]. The participants with dementia in the Hong Kong community centre generally suffer from early- to moderate-stage dementia. The involvement of the staff members and therapists in the design process can generate opportunities to empower their ownership and customise the product based on the needs of people with dementia, thereby encouraging meaningful interaction. The outcome of the co-designed interactive POF textiles will be installed in the community centre, thereby possibly evoking awareness of dementia to other stakeholders who have limited understanding of this condition.

In the initial preliminary study, the prototype of the POF textile was presented to 11 participants with dementia, five informal caregivers, two family members and two formal caregivers in a community centre's activity room (see Fig. 2). The room was dimmed and a soft POF textile sample, which has a single function of changing the colour of lighting, was displayed in front of the participants. Through observation, the engagement of the participants with dementia was possible when the formal caregivers showed them the particular gestures using the POF textile. The participants continued watching the soft lighting of the POF textile and were interested in touching the edge of this textile, which has a bright soft illumination with a rich textured form

Fig. 2 POF textile with a
participant with dementia.
Source Kim [17]

of teasel. However, they were uninterested in touching the flat surface of the POF textile.

The transition when the lighting colour changes was rapid, thereby confusing the participants when the formal caregiver asked them to select their preferred colour from the POF textile. During the activity session, the participants' interactions with the POF textile (e.g. gently touching, stroking and grabbing) were limited. An effective sensory stimulation using the interactive POF textile needs a rich and diverse range of textures, relevant visual image and colour of illumination that people with dementia can associate with. Moreover, the process of developing an interactive POF textile should be included to enhance the interactive sensory experience and attract the interest of people with dementia.

4   Conclusion

Through the review of existing research relating to design process for MSE, most of the research is limited to the development of design criteria for building MSE with the existing materials and sensory tools, particularly for people with dementia. The interactive textile, which is a fusion of embedded technology and soft textiles, can improve multisensory experience and encourage family members and caregivers to engage in meaningful interaction with people with dementia. The preliminary study indicates that the textured surface of the interactive POF textile should be further developed to attract the interest of people with dementia for their effective haptic sensory experience. Moreover, the interactive POF textiles with embedded the sensors and software interface can promote extended sensory functions. The preliminary study begins to address an issue identified with existing research into

MSE development: people with dementia's sensory preferences need to be taken into account within the design development of MSEs, particularly in relation to accessibility and usability.

Acknowledgements The researchers would like to thank the Hong Kong Sheng Kong Hui Lok Man Alice Kwok Integrated Service Centre for providing the opportunity to conduct this preliminary study.

References

1. Yu, R.: Hong Kong Jockey Club Charities, Dementia Trends: Impact of the Ageing Population and Societal Implications for Hong Kong. Impact of the Ageing Population and Societal Implications for Hong Kong. Hong Kong Jockey Club, Hong Kong (2010)
2. Lesley, C., Anke, J.: The multisensory environment (MSE) in dementia care: examining its role and quality from a user perspective. HERD: Health Environ. Res. Des. J. **10**(5), 39–51 (2016)
3. Hope, K.W., Waterman, H.A.: Using multi-sensory environments (MSEs) with people with dementia: factors impeding their use as perceived by clinical staff. Dementia **3**(1), 45–68 (2004)
4. Hope, K.: Using multi-sensory environments with older people with dementia. J. Adv. Nurs. **25**(4), 780–785 (1997)
5. Snoezelen: Multi-sensory Environment. Available from: https://www.snoezelen.info/ (2018)
6. Jakob, A., Collier, L.: Multisensory environments (MSEs) in dementia care: the role of design—an interdisciplinary research collaboration between design and health care. Design 4 Health (2013)
7. Gaudion, K.: A Designer's Approach: Exploring How Autistic Adults with Additional Learning Disabilities Experience their Home Environment. Royal College of Art, London (2015)
8. Hedman, S.: Building a Multisensory Room for Elderly Care. Multisensory Work: Interdisciplinary Approach to Multisensory Methods (2008)
9. Jakob, A., Collier, L.: Sensory enrichment for people living with dementia: increasing the benefits of multisensory environments in dementia care through design. Des. Health **1**(1), 115–133 (2017)
10. Canada, A.S.O.: Guideline for Care, in Person-Centred Care of People with Dementia Living in Care Homes. Alzheimer Society of Canada, Canada, Toronto (2011)
11. Timlin, G.A., Rysenbry, N.: Design for dementia. In: Myerson, R.G. (ed.) Improving Dining and Bedroom Environments in Care Homes. Helen Hamlyn Centre, Royal College of Art, London (2010)
12. Collier, L.: The Use of Multi-sensory Stimulation to Improve Functional Performance in Older People with Dementia: A Randomised Single Blind Trial, in Faculty of Medicine Health and Life Sciences. Southampton University (2007)
13. Jakob, A.A., Collier, L.: Sensory enrichment through textiles for people living with dementia. In: Intersections: Collaborations in Textile Design Research Conference (2017)
14. Treadaway, C., Kenning, G.: Sensor e-textiles: Person centered co-design for people with late stage dementia. Working Older People **20**(2), 76–85 (2016)
15. Treadaway, C., et al.: In the Moment: Designing for Late Stage Dementia. Design Research Society, UK (2016)
16. Kim, H.Y.: Illuminative equipment in the multi-sensory room in HKSKH Li Ka Shing Elderly Care & Attention Home in Shum Shui Po, Hong Kong (2017)
17. Kim, H.Y.: POF textile with a participant with dementia, Hong Kong (2018)

Parametric Stitching: Co-designing with Machines

Jenny Underwood ⓘ

Abstract How might we co-design with a knitting machine to support the development of wearables? Could considering knit as a parametric material system enable diverse material knowledge sets of material science, digital design and knitting textile craft and technology to come together to create a rich inter-disciplinary space. This paper offers one such approach by conceptually considering knit as a parametric material system to support creative exploration and automated efficiencies grounded in material experiences. Both the opportunities and challenges are discussed. The paper highlights that a symbiotic entangling of analogue and digital practices is needed.

Keywords Knitting · Material experience · Parametric design

1 Introduction

In the field of fashion and textiles the rise of AI technology and advanced material science challenge designers to expand their approaches to how they design. New inter-disciplinary design approaches are needed. How might the diverse knowledge domains of material science and digital design coalesce with knitting textile craft and technology? How might we co-design with a knitting machine to develop the next generation of smart wearables? This paper offers one such approach by conceptually considering knit as a parametric material system. Such an approach offers a space for creative exploration and automated efficiencies grounded in material experiences.

J. Underwood (✉)
RMIT University, Melbourne, VIC 3000, Australia
e-mail: jenny.underwood@rmit.edu.au

© Springer Nature Switzerland AG 2019
W. K. Wong (ed.), *Artificial Intelligence on Fashion and Textiles*,
Advances in Intelligent Systems and Computing 849,
https://doi.org/10.1007/978-3-319-99695-0_26

213

2 Background Context

We are on the cusp of a materials revolution. Advances in the sciences and engineering are opening up new opportunities for Fashion and Textiles industry. Concepts such as active matter [1] and a new material ecology that establish a deeper relationship between the design object and its environment are emerging [2, 3], and 3D printed voxels that potentially provide control over an object's behaviour are being developed [4].

Technology and how we engage with technology is changing. Technology interfaces are become soft and ubiquitous, with textiles transforming information technology into wearable interfaces, and haptic sensors into fibre form [5]. What new material experiences will arise from such developments? How will our clothing function, behave and be experienced?

Additionally integrated, automated and scalable approaches associated with digital tools and fabrication processes are also developing and opening up new possibilities for designers. Today, a range of new digital technology traditionally associated with engineering, industrial and automotive manufacturing processes offer enormous potential for unlimited creativity of form. Digital 3D modelling and parametric tools such as Rhino (with the plug-in Grasshopper), Maya and AutoCAD all offer ways of approaching design as a flexible system. These digital tools potentially support designers to seamlessly interface with a range of 2D and 3D fabrication technologies, such as laser cutters and 3D additive and subtractive fabrication processes such as rapid prototyping and 3D digital printing.

What do these advancements mean for design? Expanded theories of how we design are required. How can we bring the knowledge domains of material science, digital design with textile craft and technology to develop the next generation of wearables? This paper offers one such approach by considering knit as a parametric material system.

3 Parametric Approaches

The rising complexities associated with wearables requires a shift towards design being systems focused and performative, rather than artefact based and static, and to mediate emerging material and technological potential, with ecological and societal concerns.

Parametric design allows a designer to consider the relationship between elements (variables) throughout a system, so that changes in a single element distribute changes throughout the system [6]. For architecture, parametric design is made possible through sophisticated 3D CAD/CAM (computer aided design/manufacturing) software that utilizes a design feature called parametrics. Parametrics is a method of linking dimensions and variables to geometry in such a way that when the values change, the component (shape) changes as well. In this way, design modifications and creation of a family of components can be performed in remarkably quick time compared with the redrawing required by traditional CAD [6].

Parametric design has much to offer the field of 3D shape knitting leading to new understanding and generation of form and surface. This opens up the potential for a much broader range of design outcomes that sit outside the boundaries of conventional garment typologies. Such processes also makes possible for the complete integration and automation of design through to fabrication, as well as for both pre and post-production marketing communication. A design could be evaluated by a team via digital 3D simulation, allowing for real time feedback loops within the design process. Rather than having to redraw a design, variables within the system are merely adjusted.

For fashion design ease of transfer of 3D parametric modelling tools to 3D printing offers significant potential to expand existing design methods. For example the work of designer Iris van Herpen's 3D printed Escapism Couture collection shows the potential for how new design tools and technologies contribute to an emerging range of aesthetics and for the precise control of material performance [3]. However, while 3d modelling can easily dialogue with 3D printers, such tools do not easily connect to conventional textile technology, such as an industrial knitting machine.

4 Knit as a Parametric Material System

Weft knitting is fundamentally a mathematical process. It is a highly organized and logical with a set of rules that enable infinite potentialities in the way a single thread (yarn) can form a sequence of interlocking stitches. It is through the repetition of the sequence of stitches that enable the 2D surface (fabric) and 3D form to systematically emerge.

With this inherent logic of knitting in mind, applying an understanding of basic principles of parametric design to knitting opens up considerable design opportunities. Consider a simple knitted form of a triangular segment generated by suspended stitches (holding stitches) (Fig. 1).

How the form is knitted is determined by a number of variables and potential constraints (or fixed variables). By understanding these variables and constraints, greater control and 3D form generation is possible through the adjustment of one or more variables. As demonstrated through the identification the four variables of a triangular segment, each responsible for specific performance qualities of the artefact, a diverse range of shape generation is achieved using the suspended stitches technique. From a flat triangular segment, a range of complex 3D parabola-like and spiral structures emerge as possible. Altering any one or all of these four variables ultimately affects the overall size and structural dimensions of the form, as well as allowing for a wide range of other cone-like structures such as hourglass and funnels to be formed. Additional variables and constraints could be added to the parametric model and the complexity increased. Such variables might relate to the fibre, yarn, colour and surface stitch architecture. The basic 3D knitted forms presented in Fig. 1

Key variables include:

(1) The shaping segment type

(2) The number of shaping segments

(3) The number of courses between the shaping segments

(4) The number of stitches

Fig. 1 The applying the principles of parametric design to knitting [11]

represent the base building blocks. By combining these base blocks, more complex outcomes are possible.

Extending the form-finding potential of knit further, to incorporate other performance and behaviour criteria, a material system emerges that can incrementally deal with greater complexity. Such an approach importantly starts to provide a space for inter-disciplinary dialogues, as the system can be tinkered with and options explored. A type of co-design between researchers and with the knitting machine emerges. This could lead to a space where both technical and poetic concerns are evaluated in real time, and is open enough for end users can be brought into the early prototyping design stage to inform and enrich the process.

While systems are emerging, to date they are still somewhat rudimentary and problematic. When the design is transferred to the specialized software of an industrial of knitting machine much hand coding, data cleaning, corrections and additional data input specific to the machinery is required. The process is far from automated and ultimately becomes heavily dependent on the skill of the knit technician to translate the design into a knitting logic.

Research such as the work led by Mette Rasgaard Thomsen at CITA Copenhagen does provide cues for the better integration of design and fabrication technologies and for how technologies such as sensors and simulations can be integrated into a material knit system [7] through inter-disciplinary approaches. Within the work, the relationship between the modelling and CNC knitting machinery becomes important in building a material knowledge of making. For example the Listener project developed in collaboration with Shenkar College of Engineering and Design which explored the "link between standard architectural design environments (Rhino and Grasshopper), CNC knitting machine (Stoll) and simple computational steering (Arduino)" [8].

In the area of 3D virtual garment design, while programs such as Marvellous Designer, and Opti Tex are also developing sophisticated tools they are still limited in scope, only simulating textiles with highly simplified models of behaviour. The focus of these simulations is to understand the visual quality of the resulting textiles rather than their performative behaviour. They do not directly interface with the textile machinery. The material knowledge is predefined through software libraries intended for visualization. This results in a limited scope for genuine material engagement.

Knitting machinery companies such as Shima Seiki and Stoll have developed tools that integrate the design, visualization and virtual prototyping of textiles. These rely on extensive material libraries of yarns, stitch structures and patterns and effectively makes the fabric a flat 2D surface. While the virtual 2D fabric can be draped across an avatar, with a pattern mesh being laid across a specified garment type. Design options are limited to predetermined garment types, thus limiting the designer's involvement and effectively ignoring the designer's material experience.

More recently is the research out of Carnegie Mellon University that shows the development of a computational approach that can transform 3D meshes, created by traditional modelling programs, directly into instructions for a computer-controlled knitting machine [9]. They have been able to demonstrate this approach using a wide range of 3D meshes.

Such research highlights the complexities and challenges that presently exist between bringing design and fabrication technologies together. As more advanced materials are integrated such as sensors these systems become even more complex. Therefore, a parametric material system that can understand the unique and inherent logic of knitting could provide a way forward to improve the interface.

Inherent in the development of a parametric material system is textile craft knowledge associated with the material experience of knit. Knowing how a knitted fabric behaviours (technically and poetically) must inform and constrain the emerging digital system. The interplay between the analogue and digital therefore becomes critical. There is a need to ensure that material information is embedded into the digital design models, so that the conditions of material constraints are able to be understood. Otherwise this digital culture risks being abstract ideas isolated from cultural context and embodied cognition [10].

Central to this understanding is knitting's materiality. For example each unique combination of variables such as the; fibre, yarn, stitch architecture, stitch quality (tension), stretch (be it in the wale, course or bias direction), weight, drapeability, machinery and so on all impact on the knitted textiles performance qualities. It requires an embodied craft based knowledge of knitting. By incorporating such variables within a system, the potential for meaningful engagement increases, but it is far from precise. This is because of the highly flexible nature of knitting; these values cannot be considered absolute. However, for a parametric model to be useful, accurate measurements, performance qualities and fabric variables need to be understood. Therefore what is required is a two way flow of information/data. The analogue and digital must co-support each other and ultimately inform each other. In addition, there must be a flow of knowledge between the different discipline domains. Embodied textile craft must work alongside material science and textile technology, in more connected ways.

5   Conclusion

A parametric material system approach has much to offer fashion and textile as a way to develop new modes of practices with improved efficiencies. In the field of wearables this could expand the potential for creative and socially resonant design outcomes to emerge. By taking a systems approach and focusing on the relationship between variables throughout the system, rather than specific values, changes how design is approached and the potential for new innovative forms to emerge outside the boundaries of conventional garment typologies and material experiences. Expanding such an approach to include knowledge of material experience such as how a knitted fabric behaviours(technically and poetically) allows for an inter-disciplinary space to emerge and for a flow of information between the digital and the physical. This offers the potential for a holistic system to emerge that might mediate material and technological potential, with ecological and societal concerns. In addition a parametric material approach could make possible the complete integration of design, communication and fabrication processes, enabling real time feedback loops to be built into the design process. But such a system requires complex diverse inter-disciplinary knowledge and skills. It requires an understanding of material design knowledge, tacit crafts, advanced technology and automated fabrication processes.

References

1. Tibbits, S. (ed.): Active Matter. MIT press, Cambridge, MA (2017)
2. Myers, W., Antonelli, P.: BioDesign: Nature Science Creativity. Thames and Hudson, London (2014)
3. Oxman, N., Rosenberg, J.: Material-based design computation: an inquiry into digital simulation of physical material properties as design generators. International Journal of Architectural Computing (IJAC) 5(1), 26–44 (2007)
4. Lison, H., Kurman, M.: Fabricated: The New World of 3d Printing. Wiley, Hoboken (2013)
5. Quinn, B.: Textile Futures. Berg, London, UK (2010)
6. Hernandez, C.R.B.: Thinking parametric design: introducing parametric Gaudi. Des. Stud. **27**, 309–324 (2006)
7. Ramsgaard Thomsen, M., Tamke, M., Karmon, A., Underwood, J., Gengnagel, C., Stranghöner, N., Uhlemann, J.: Knit as bespoke material practice for architecture. In: Velikov, K., Ahlquist, S., del Campo, M., Thün, G. (eds.) Proceedings of the 36th Annual Conference of the Association for Computer Aided Design in Architecture (ACADIA 2016), ACADIA 2016: Posthuman Frontiers: Data, Designers, and Cognitive Machines, Netherlands, pp. 280–289 (2016)
8. Ramsgaard Thomsen, M., Bech, K.: Textile Logic for a soft space. The Royal Danish Academy of Fine Arts Schools of Architecture, Design and Conservation School of Architecture, Copenhagen (2011)
9. Naraynan, V., Hodgins, J., Coros, S., McCann, J.: Automatic machine knitting of 3D meshes. ACM Trans. Graphics **1**(1), 35 (2018)

10. Gallese, V., Lakoff, G.: The brain's concepts: the role of the sensory-motor system in conceptual knowledge. Cognitive Neuropsychol. **22**(3–4), 455–479 (2005)
11. Underwood, J.: The Design of 3D Shape Knitted Preforms. PhD, RMIT University, Melbourne, Australia. Retrieved from https://researchbank.rmit.edu.au/eserv/rmit:6130/Underwood.pdf (2010)

Live: Scape BLOOM: Connecting Smart Fashion to the IoT Ecology

Caroline McMillan

Abstract *Live: scape BLOOM* is an Internet of Things connected dress. This paper outlines methods for constructing smart fashion for an IoT ecosystem within a framework of new materialism (Barrett and Bolt in Carnal knowledge: towards a 'new materialism' through the arts. I.B. Taurus, London, 2013 [1]). I argue the potential for tangible forms of digital entities, such as data, can be observed through material dynamism. Traditional fashion couture techniques converse with digital tools, employing makerism to create the robotic textile embellishment with intuitive real-time environmental data streaming. Data is experienced kinetically through textile surface embellishment to extend its dimensionality. By foregrounding aesthetics and material development in the design inquiry, valuable insights for tangible forms of digital entities are revealed. The technologies of an IoT ecology are used to reconfigure conceptual shifts for user-curated data programs in smart fashion.

Keywords Internet of things · Smart fashion · Makerism

1 Introduction

The Internet of Things is a networked ecosystem that aims to connect everyday objects; virtual, physical and material. Applications for wearable concepts within this ecology are emerging as a broad spectrum of future research directions. However, the impact of the IoT on contemporary smart fashion design has been under-represented in literature and is not yet clear. This paper describes *Live: scape BLOOM*, which responds to issues about our growing collection and interaction with data. The aim was to develop a dress with decorative textile embellishment capable of connecting wirelessly to real-time data flows to enable pattern changes over time through kinetic motion. Exploring wireless connection to investigate the use of data flow materiality

C. McMillan (✉)
RMIT University, Melbourne, VIC 3000, Australia
e-mail: cazmcmillan@hotmail.com
URL: http://orcid.org/0000-0001-9329-9838

© Springer Nature Switzerland AG 2019
W. K. Wong (ed.), *Artificial Intelligence on Fashion and Textiles*,
Advances in Intelligent Systems and Computing 849,
https://doi.org/10.1007/978-3-319-99695-0_27

221

is an opportunity for additional dimension for smart fashion textile embellishment. This project addresses the combined physical and digital material behavioural characteristics of textile, hardware, software and real-time data through the design of garments that employ the latest in mobile technologies. Real-time data flows are used to create aesthetic affect. The focus of this investigation is the source of where the data comes from in a wearable context and how do we want to use significant sources of data? How will it alter both our interaction with data and ideas of materiality? *Live: scape BLOOM* compels us to imagine structural and dimensional possibilities for decorative surface design on a smart fashion garment.

2 Background

Garments exploring data visualization of interactive technologies vary. Sensor-driven garments such as the *Diffus Design Climate Dress* (2009) collect environmental data [3]. Through a framework of soft computation, the garment uses textiles to visualize climatic data as LED light patterns, politically engaged as an active statement responding to debates about environmental issues. Valérie Lamontagne's *Sky Dress* [9], one of a trio of dresses in an installation called Peau d'âne, wirelessly connects to meteorological data to trigger a voluminous, shape-changing silhouette. Aesthetics and poetic expression are the driving force behind Lamontagne's concepts, inspired by natural weather patterns of wind. Freire and Bruckner propose that mindful, user-curated data consumption garners autonomy and wearer empowerment [5]. *Embodisuit* (2017) is a haptic, IoT-connected garment that reverses the notion of the quantified self. Modules transmit incoming data flows, ubiquitously vibrating alerts on the wearer's skin. It is important to reassess the potential for wearable computational interactions, beyond notifications and alerts. Galloway [6] argues for being playfully mobile in the framework of wireless and ubiquitous technologies. To further demonstrate this concept of playful critical analysis, Kobakant celebrate the technologies with which they engage. *Smart Rituals* (2016) involved a series of IoT-connected, intelligent garments where theatrically dressed, networked citizens build a better smart city, using gesture-triggered sensors to collect and process ambient environmental data [8]. Quinn [10] posits the role of the wearer is shifting to a systems user and will create significant changes. It is right to question what will be the nature and function of our clothes, and what behavioural shifts will ensue.

3 Live: Scape BLOOM

Live: scape BLOOM is an IoT-connected dress whose floral embellishment changes mode over time in response to real-time, meteorological data streams. To reconfigure the contemporary wearables approach, the idea to use a sensor layer for *Live: scape BLOOM* to collect ambient environmental data was abandoned in favour of connecting wirelessly to open source meteorological data. This preference of connecting to a remote, live data feed allowed the possibility for a wearer to choose a curated program of data flow meaningful to them, as opposed to an automated, quantifiable accumulation of personal biometric data. With its emphasis on remotely sourced data, rather than personally harvested biometric data, this intention critically reverses the paradigm of the Quantified Self, the practice of self-tracking information [13]. Instead, the garment draws upon a program of user-curated data. *Live: scape BLOOM* integrates traditional and unusual textiles, electronics and the latest developments in mobile technologies to extend the floral motif (Fig. 1). It aims to represent the aesthetic qualities of both the physical, behavioural characteristics of textile and the ephemeral, changeable and transitory material qualities of the data sourced. Meteorological data is not stable. It is in a constant state of flux. In this manner, I wanted to observe the real-time data feed, experience its material qualities and illustrate the ephemeral qualities of a digital entity. Fluctuations of wind speed were data mapped, coded to reflect the slow, gentle, and at times unpredictable and forceful. With the addition of electronic mediation, in performance *Live: scape BLOOM* alternates between symbolic representation and agential 'aliveness' of the flowers and the invisible, agential force behind its movement, the wind, blurring these distinctions. The bio-inspired aesthetic of the kinetic embellishment references the behavioural practice of wind patterns and the effect of these patterns on flowers in nature.

Why flowers? "To the Nlaka' pmx, who lived in the interior of southern British Columbia, flowers were "the valuables of the earth. If they were destroyed or plucked too ruthlessly, the Earth sorrowed or cried [7]." In addition to cultural and historical significance, as a motif flowers are commonly used as a decorative embellishment in fashion couture [2]. Lagerfeld describes the inspiration behind the Spring 2015 haute couture collection for Chanel, "but these are flowers God had forgotten to create. They don't really exist these kinds of flowers" [4]. In *Live: scape BLOOM*

Fig. 1 Shows a still image from a documentation film of the Live: scape BLOOM proof of prototype, focusing on the garment's wifi-connected, robotic, decorative embellishment

traditional fashion couture techniques are fused with robotic mechanisms and code to create a novel representation of the flower motif, as well as demonstrate emerging research in networked mobile technologies. The kinetic flowers leave an invisible, performative wake [6], the added dimensionality of the kinetic flower embellishment occupies more space through movement, form, volume and presence.

However, the movement ultimately becomes illustrative though the garments hardware and software embellishment system, creating a movement nature never intended, a gesture of affect that alludes to aliveness and garners an affective response. "This dress is flirting with me," states a viewer at the edges and in between exhibition (A. Petidis, personal communication, March 7, 2018). The nuanced qualities afforded in these responses to this dress can be leveraged to unearth cultural relevance, affect, social engagement, and critical reactions to a paradigm of technological functionality that has dominated contemporary wearables design.

4 How It Works

The research focuses on integrating wireless communication technology into a main-stream fashion dress using traditional textile embellishment techniques, to enhance functionality and extend aesthetic possibilities. This aim is achieved through a deep understanding of the potential of decorative textile methods, construction and material qualities. Also, skills acquired through makerism from electronics and robotics are transferred back into my fashion wearables design practice. *Live: scape BLOOM* integrates traditional fashion construction techniques and sophisticated technologies with unusual materials for networked, pattern change embellishment. It uses conventional textile fabrics, jewellery beads and trims, electronics components, a wifi development kit, servo-motors and custom software. With the aim to seamlessly integrate technology and user interplay with data, into the tradition of decorative, textile surface design for fashion haute couture.

5 Construction Process

Live: scape BLOOM is constructed out of a variety of textile materials and takes the form of a mainstream fashion dress, using electronic components. The aim was to construct the robotic flowers using only textile, firstly for computational ubiquity, to minimize the look of technological components and secondly, to flip the paradigm of technological functionality and instead foreground aesthetics. With material properties at the forefront of the investigation, behaviours of softness, lightness, drape and biomimetic movement are leveraged. The base fabric is vegetable-tanned leather, providing a structured foundation to house the hardware and support the placement of the robotic flowers around the neckline. A delicate lace pattern is laser cut using digital fabrication to create a contemporary ready-to-wear fashion look. The embellished floral appliqué is both crafted by hand and machine sewed, using silk organza and nylon, laser cut plexiglass, foam sheets, metal jewellery wires, beads and feathers. Small, lightweight servo-motors are powered by two battery packs, which are connected to the microcontroller to animate the robotic flowers with real-time data, via wireless cloud computing. The electrical wiring between the flowers and the circuitry is encased in a protective, channelled, nylon binding to increase comfort. The robotic embellishment is then hand draped, fastened and hand sewn onto the garment bodice.

6 Wireless Processing, Robotics and Cloud Computing

Initially, it was necessary to explore many different fabrication options to integrate the textile, robotics, software and electronic hardware components. Material weight and behavioural characteristics influenced the performance of the kinetic, robotic flowers, drape and form of the garment and suitability for digital fabrication methods such as laser cutting. Chromium-free, vegetable-tanned leather and weight-less natural textile materials such as silk organza were combined with textile robotic mechanisms, modified circuitry and connective technologies to create a garment with the desired aesthetic outcome of a mainstream fashion dress. Such materials responded to a traditional textile language of form through sculptural dimension on the body.

The robotic flower mechanism required an intensive process of reflection through action. Digital fabrication conversed with traditional embellishment and low-tech methods iteratively to achieve an artisanal textile couture outcome (Fig. 2), with an animated, lifelike enchantment powered by a synthetic operation. Drawing on notions of reflective [12] and generative [11] design process, proto-typical iterations of the robotic flowers included; experimenting with digital fabrication methods trialling 3D modelling software, 3D printing and laser cutting techniques by printing a robotic flower casing (Fig. 2a); paper constructed prototypes compared the action of linear actuators and servo-motors, simulating hydraulic and rotational mechanisms (Fig. 2b, c); and employing hand crafted, fashion haute couture-derived embellishment approaches to constructing a textilized, robotic flower (Fig. 2d). Due to the power inefficiency and limited response of shape memory alloy wire, robotics were used to facilitate the transformative kinetic movement for Live: scape BLOOM dress. Coded and programmable, robotic mechanisms facilitate greater operational scope and impart a lifelike quality to the dynamic flowers. As a classically trained fashion designer familiar with reflective and generative approaches, making these connections with the processes of developing robotics and software scaffolded a process where key technical skills could be transferred back into traditional methods of textile decorative embellishment couturiers, drawing from diverse material cultures to create the hybrid design. For ubiquitous online connectivity, to operate independently of a wifi network, the garment is equipped with a SIM card in the microcontroller and activates a data plan subscription. The basis is then established for the IoT action layer, the event detection and visualization of real-time information, to animate a

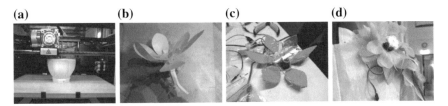

Fig. 2 **a–d** In the maker lab, reflective design iterations in the prototyping process of constructing the robotic flowers

digital behavioural interaction with the physical world. The system architecture was built using a wifi development kit, consisting of a microcontroller, software platform, antenna and cellular network. Wirelessly connecting the garment for cloud computing enabled significant software and data processing, reducing hardware components and wires.

7 Conclusion

This research provides a model for IoT-connected, smart fashion. It reveals broader aesthetic potential and wearable applications for the future of networked garments in an IoT ecosystem, which has been under-represented in literature. It foregrounds material development for aesthetics in fashion design by responding to mainstream fashion trends, while investigating ephemeral, invisible forms of technological entities such as data where movement traces can be witnessed over time. Textile material characteristics determine aesthetic properties, hardware configuration and real-time data processed by cloud computing, electronically integrating movement into the garment's decorative surface display. The use of robotic mechanisms and wireless technology to connect to a user-curated data source enables an array of dynamic patterns, extending the use of traditional textile embellishment. Kinetic and form shifting fashion are a big part of the future of smart fashion for personalized display and interface possibilities, how we will interact with data. However, programming must consider the frequency of the IoT data events to create a discernible movement for poetic gestures.

The opportunities for wearable, IoT-connected forms as smart fashion are diverse. Nanophotonics for artificial intelligence can extend the concepts: both to create wearable applications of novel material development and to radically improve smart fashion hardware and battery power supply constraints for increased, more resilient wearability. Concepts can be further developed for the future of health, communications and lifestyle as networked, physically adaptive visualizations, responsive to a user-customised program of use. Live: scape BLOOM encourages discussion about how to aesthetically respond to and interact with data flows on the human body through smart fashion, and prompts discourse on upcoming social and cultural trends issues that will emerge from an increasingly connected, networked ecosystem.

References

1. Barrett, E., Bolt, B.: Carnal Knowledge: Towards a 'New Materialism' Through the Arts. I.B. Taurus, London (2013)
2. Chanel: Making-of the CHANEL Spring-Summer 2015 Haute Couture Collection, https://www.youtube.com/watch?v=3cDhTzoMj_s. Last accessed 20 Apr 2018
3. Climate Dress: http://www.diffus.dk/climate-dress/. Last accessed 29 Apr 2018
4. Frankel, S.: The Story Behind Chanel Couture, http://www.anothermag.com/fashion-beauty/4 300/the-story-behind-chanel-couture-spring-2015. Last accessed 20 Apr 2018
5. Freire, R.: Embodisuit, http://www.rachelfreire.com/embodisuit/. Last accessed 12 Mar 2018
6. Galloway, A.: Playful mobilities: ubiquitous computing in the city. In: Alternative Mobility Futures Conference, pp. 9–11 (2004)
7. Gessert, G.: Green Light: Toward an Art of Evolution. MIT Press, Cambridge (2012)
8. KOBAKANT: Workshops, Smart Rituals, http://www.kobakant.at/DIY/?p=6204. Last accessed 11 Mar 2018
9. Lamontagne, Valérie: PEAU D'ÂNE (2005)
10. Quinn, B.: Techno Fashion. Berg, Oxford (2002)
11. Sanders, E., Stappers, P.: Probes, toolkits and prototypes: three approaches to making in code-signing. CoDesign **10**(1), 5–14 (2014)
12. Schön, D.A.: The Reflective Practitioner: How Professionals Think in Action. Basic books, New York, USA (1984)
13. Swan, M.: The quantified self: fundamental disruption in big data science and biological discovery. Big Data **1**(2), 85–99 (2013)

Traps in Multisource Heterogeneous Big Data Processing

Yan Liu

Abstract The importance of big data values and application efforts has reached a universal consensus in most fields. While because of the model difference of data storage, computing and analysis, the big data processing performance and big data values show greatly uneven in different scenarios. In this paper, we analyze the traps which may greatly impact big data processing results, and give our suggestions to solve these problems for the multisource and heterogeneous characteristics of big data.

Keywords Fault tolerance · Data credibility · Data fusion

1 Introduction

Big data, which shows the great potential and has been considered to be a digital asset, is increasingly attracting more and more attention. They are being widely used in risk prevention, anti-fraud, consumer portraits, management decision, etc., in different industries. The space we live in is filled with big data. The network enlarges the scope of data acquisition, and all kinds of high-precision equipment increase the data's sampling density. Big data helps human to think jumping out of the world no matter from a broader perspective or from a more microscopic perspective. The more concentrated the knowledge distribution becomes, the higher the density value of big data will be. Although the application of big data is very meaningful, there are many challenges in dealing with big data [1].

Big data is very different from traditional data in terms of data structure, data distribution, and data volume, etc [2]. Therefore, research on big data processing methods is a hot and difficult topic [3]. As shown in Fig. 1, due to the extensive sources of big data and the lack of data standards, existing methods in data collection and data fusion are not able to achieve promising results, and custom development is quite

Y. Liu (✉)
Taikang Insurance Group, Beijing, China
e-mail: lyan2006@gmail.com

© Springer Nature Switzerland AG 2019
W. K. Wong (ed.), *Artificial Intelligence on Fashion and Textiles*,
Advances in Intelligent Systems and Computing 849,
https://doi.org/10.1007/978-3-319-99695-0_28

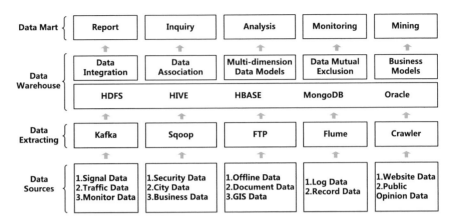

Fig. 1 Big data flowing framework

necessary. In addition, from the view point of data volumes, over 85% of big data are semi-structured and unstructured data. These data cannot be stored and processed by traditional relation database. So now there are many NO-SQL databases, such as MongoDB, HBase, Neo4j, and so on. These databases often support open source computing frameworks such as Hadoop. Hadoop, Spark and Storm are the three most popular big data processing frameworks [4, 5]. They are normally applied to three different scenarios: mass batch processing, memory batch processing and streaming processing respectively.

The knowledge involved in multisource heterogeneous big data processing is very broad. In this paper, we mainly focus on the discussion of four thorny problems in the following sections, and give our suggestions to solve each problem.

2 Unstructured Data Processing

The typical unstructured data includes video, image, non-layout text, etc. [6] The treatment between unstructured data and structured data is greatly different. For unstructured data, structuring is a necessary step. In the data center of safe cities and smart cities, video data is the scale-largest and most typical big data. Although Hadoop has been widely used to perform big data processing, it has few cases of video processing. The reason may be the gap between distributed computing and data analysis algorithms.

MapReduce, as the core computing unit of Hadoop, implements distributed parallel computing based on data partitioning. And the maximum file size that it can handle is 256 MB. Therefore, for video stream data, the first step is to divide the stream into blocks. For video stream partition, there are three methods: frame-based partition, shot-based partition, and object-based partition.

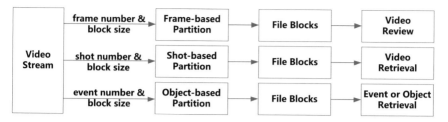

Fig. 2 The process of video stream partition

(1) Frame-based partition methods control file size based on frame number of res-
     olution. This method can accurately handle the file size, but it breaks the conti-
     nuity of video content.
(2) Shot-based partition methods control file size based on the video shot cut. Sim-
     ilar video shots will be divided into the same file.
(3) Object-based partition methods control file size based on the object tracking.
     The same objects or events are divided into the same file.

As shown in Fig. 2, different file partition methods are designed for different
application scenarios. Frame-based method can achieve very high file size accuracy
but does not consider the frame contents, so it can be used as a regular file partition
method. Shot-based method considers the continuity of video content's changing, so
the files it cuts can be used for video retrieval and video compression. Object-based
method considers the moving objects' changing in video, so the files it cuts can be
used for object retrieval and event detection. In short, unstructured data processing
should design parallel processing framework from the perspective of data analysis
algorithm as much as possible, and this idea runs through the whole data processing
procedure.

3   File Fault Tolerance Storage

Frequent file reading and writing may damage files, especially small files. And the
most common manifestation of this damage is that the file becomes unreadable. To
explain this problem clearly, we choose the most popular storage systems, HDFS
and FAST-DFS, as examples for illustration. HDFS solves the problem of large file
storage and suit sequential read type applications. FAST-DFS solves the problem of
large capacity storage and keeps balancing. This method is especially suitable for
small and medium files. In both of these two systems, damaged files can be recovered
by backup strategies but no measures are provided to prevent or reduce file damages
[7]. In the next, we shall show one file storage solution to enhance the fault tolerance.

We merge and store the files according to the file size that MapReduce (optional)
can handle. Each file is stored as a line, and the beginning field of each line labels the
meta-data, such as file source, size, format, time, etc. These meta-data and data body

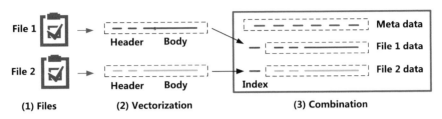

Fig. 3 The process of file combining storage

Fig. 4 An example of data conflicts between different data sources

form the key-value family pair, where the definition of family is similar to HBASE. Thus reading and writing file operations will be replaced by visiting different text lines. Even though some bytes in the line are wrong or damaged, there is less influence on file reading and writing. This solution is much suitable for large-scale small file storage (Fig. 3).

4   Multisource Data Credible Computing

The diversity of data source is one of the most important characteristics of big data. There are lots of conflicts among data from different sources. As illustrated in Fig. 4, someone's mobile number is 135222** in data source A, but 138666** in data source B. Or the owner's name of the same mobile number is different in different data sources. Therefore, how to measure the credibility from different sources is extremely necessary.

The credibility of the same data is different in different data sources. In high-confidence data sources, not all the data are trustworthy. Similarly, in low-confidence data sources, some data also have high confidence. For example, in the Ministry of Public Security database, the credibility of person's name is very high, but the credibility of the mobile number is very low. On the contrary, in the express database, the credibility of person's name is very low, but the credibility of mobile number is very high. Here we give some empirical principles for evaluating the credibility of different data sources.

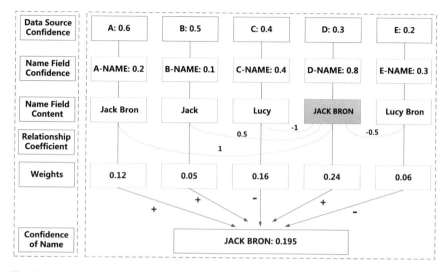

Fig. 5 An example of multisource data fusion, where A, B, C, D, E symbolize different data sources, the last confidence of name is $0.195 = 0.12 * 1 + 0.05 * 0.5 - 0.16 * 1 + 0.24 * 1 - 0.06 * 0.5$

(1) For the relatively fixed data in the long term, such as person's name, ID number, gender, the data from government and financial institutions are more trustworthy.
(2) For the communication data, such as address, contact, the data from the courier companies and Internet companies are more trustworthy.
(3) For the locations and wealth level data, such as GPRS signal, mobile signal, the data from communication companies and passenger transport companies have higher confidence.
(4) Data sources and its inner data fields should be calibrated with confidence separately before data cleaning and fusing.

5 Multisource Data Fusion Analysis

Since the credibility of different data sources is different, there may be contradiction between the data of different sources, it is very necessary to design targeted data fusion algorithms. In this section, we illustrate the correlation of data sources and fields, and present the importance of data distribution for data accuracy analysis.

Data Correlation. Data from different sources may be positive or negative for data fusion analysis. The positive correlation data will strengthen the confidence, and conversely, the negative correlation data will weaken the confidence. As illustrated in Fig. 5, there are five data sources, symbolized as A, B, C, D, E separately, where a person's name is recorded as Jack Bron in A, Jack in B, Lucy in C, JACK BRON in D, Lucy Bron in E. Here we need to infer the confidence of the person's name.

First the name field with the biggest confidence is chosen as a candidate. Here JACK BRON with confidence 0.8 in data source D is chosen as the hypothetic name. Then the relationship coefficient of different name fields are ascertained by comparing name filed. It is easy to know that A and D are completely positive (the coefficient is 1), A and B are semi-positive (the coefficient is 0.5), A and C are completely negative (the coefficient is -1), A and E are semi-negative (the coefficient is -0.5). The weights of different data sources can be obtained by multiplying data source confidence with name filed confidence. At last, the confidence of name JACK BRON is computed by fusing weights of different data sources.

Data Distribution. The data distribution determines the choice of the weight probability density function. Thus determining the distribution of data is the premise of data analysis. The same batch of data may have different confidence weights at different distribution points. For example, the peak of express service usually appears at 10 AM and 4 PM. There is no special difference between the single express data, but on the whole, the business density distribution is presented as two straw hat shapes in the time dimension. The confidence weights of different data can be represented by some distribution functions, such as the Gaussian probability density distribution function. Thus different data have different weights during the fusing procedure. This processing way can enhance the weights of key data in the overall analysis, so it promotes analytic conclusions closer to the truth.

Determining the relationship between different data sources is a key step of multisource data analysis. There is a dependency or conflict between data from different sources. This phenomenon is more pronounced in process? and behavior data. For example, someone appears in a certain place in data source A, thus at the same time he can never appear in another place in data source B, or he must also appear in a certain place in data source C. This is a kind of joint distribution question among data sources. To deal with such a problem, we first need to sort out the relationship of data sources, and then clarify the association between data fields. Strip off those independently distributed data sources or fields, and determine their independent distribution density function. For the correlated data sources, determine their joint distribution density function.

6   Conclusion

In this paper, we mainly talked about several problems in multisource heterogeneous big data processing. When facing different data contents and data structures, the manifestations of these problems are different and also have no readymade solutions. For each problem, the processing suggestions we give are not unique and fixed. We aim to provide a processing idea for data researchers. For multisource heterogeneous data processing, it is necessary to weight the balance between big data's ETL (Extracting, Transformation, Loading), storage, computing, network communicating, and fusing analysis algorithms so as to reduce resource consuming.

References

1. Lazer, D., Kennedy, R., King, G., Vespignani, A.: The parable of Google flu: Traps in big data analysis. Science **343**, 1203–1205 (2014)
2. Storey, V.C., Song, I.I.-Y.: Big data technologies and management: what conceptual modeling can do. Data Knowl. Eng. **108**, 50–67 (2017)
3. Kwon, O., Lee, N., Shin, B.: Data quality management, data usage experience and acquisition intention of big data analytics. Int. J. Inf. Manage. **34**, 387–394 (2014)
4. Basanta-Val, Pablo: An efficient industrial big-data engine. IEEE Trans. Ind. Inf. **14**, 1361–1369 (2017)
5. Rui, H., Lizy, K., Jianfeng, Z.: Benchmarking big data systems: A review. IEEE Trans. Serv. Comput. **11**(3), 1–17 (2017)
6. Xing, E.P., Qirong, Ho, Wei, Dai, Kru, Kim Jin, Jinliang, Wei, Seunghak, Lee, et al.: Petuum: A new platform for distributed machine learning on big data. IEEE Trans. Big Data **1**(2), 49–67 (2015)
7. Aishwarya G., Ramnatthan A., et al.: Redundancy does not imply fault tolerance: analysis of distributed storage reactions to single errors and corruptions. In: Proceedings of 15th USENIX Conference on File and Storage Technologies, pp. 149–165 (2017)

Convolutional Neural Networks for Finance Image Classification

Xingjie Zhu, Yan Liu, Xingwang Liu and Chi Li

Abstract In recent years, deep convolutional neural networks have demonstrated excellent performance on visual tasks, such as image classification. Based on the deep architecture, this paper designs a new method to handle an automatic financial recognition problem, namely to recognize the value-added tax (VAT) invoices, finance-related documents, and other finance-related images. The proposed method consists of three steps: first, image preprocessing will be performed on the original image and the augmented image will be separated into four patches for further processing; thus the obtained image patches will be the input of a deep convolutional neural model for the training purpose; at the final step, we use the four predications which obtained from the previous step to determine the final categorizes. Our experimental result shows that this method can conduct finance image classification with high performance.

Keywords Image classification · Segmentation · Neural networks

1 Introduction

Image classification, which can be defined as the task of categorizing images into one of several predefined classes, is often a prerequisite step towards finance images filing tasks, such as information extraction using vision, text recognition, etc. Although these tasks can be easily accomplished by humans, it is much more challenging for an automated system. Some of the encountered complications include the variability of viewpoint-dependent object and high similarity of different object types. Traditionally, a two-stage approach was used to solve the classification problem [1–3]. The commonly used image-based feature extraction techniques include intensity histograms, filter-based features and the popular scale-invariant feature transform (SIFT) [4] and local binary patterns (LBP) [5, 6]. The extracted feature vectors are

X. Zhu · Y. Liu (✉) · X. Liu · C. Li
Taikang Insurance Group, Beijing, China
e-mail: lyan2006@gmail.com

© Springer Nature Switzerland AG 2019
W. K. Wong (ed.), *Artificial Intelligence on Fashion and Textiles*,
Advances in Intelligent Systems and Computing 849,
https://doi.org/10.1007/978-3-319-99695-0_29

normally used to train a classification model, e.g., the support vector machine (SVM) [2, 7] and sparse representation [3, 8] to obtain the image categories. A major drawback is that these methods cannot well cope with the classification problem under the above-mentioned complicate situations.

Artificial Neural Network (ANN) has been studied for many years to solve complex classification problems such as image classification [9]. Deep learning models that exploit multiple layers of nonlinear information processing have shown promising performance to overcome the difficulties for feature extraction, feature transformation, and classification. In recent years, convolutional neural networks (CNNs) [10, 11] have become the leading architecture for most of the image recognition, classification, and detection tasks. Several advances such as the first GPU implementation [12] and the first application of maximum pooling (max pooling) for DCNNs [13] have all contributed to their recent popularity. Moreover, when the content and structure in finance images are unconstrained, pre-defined features may not be able to capture all variations of a particular class.

Our aim of this study is to find a good method to classify different categories of finance images. However, for the finance images, due to high visual variation within the same class and high visual similarity between different classes, it is very challenging to achieve accurate classification. Such difficulties can be seen from Fig. 1. The main problem is thus how to design a highly discriminative feature set to effectively handle the within-class variation and between-class similarity. In this paper, we use a common CNN network for finance image patch classification. Rather than defining a set of features manually [4, 6], we designed a fully automatic neural-based machine learning framework to extract discriminative features from training samples and perform classification at the same time. Experimental results as shown in Sect. 4 validates the effectiveness of the proposed method.

The rest of this paper is organized as follows. In Sect. 2, we describe the details of image processing before training step, and introduce the proposed method to obtain the final prediction after receiving the probability from the training model. We evaluate the performance of the proposed algorithm in Sect. 3 and conclude the paper in Sect. 4.

2   The Proposed Method

The finance images are often contains two parts, which through use different finance paste into the white papers to obtained. This special forms means it is difficult to achieve accurate classification by common method. Besides, large input dimension not only costs more computation resources but also leads to greater chance of overfitting. Therefore, in order to solve this problem, our method mainly includes three steps (as Fig. 2 described): first, we obtain the segmented images derived from the input finance images, which can obtain the true finance images from the input images; second, we use the segmented images to train a deep neural network model, which

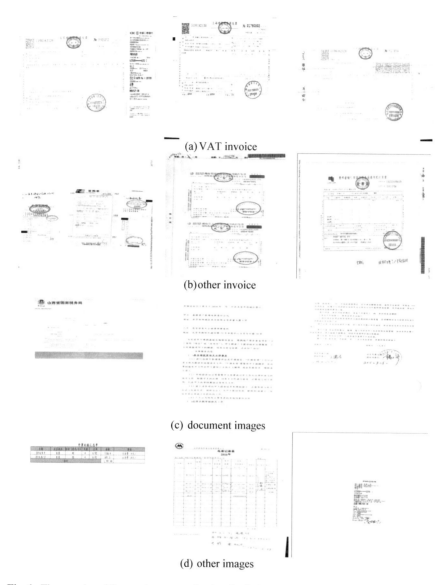

(a) VAT invoice

(b) other invoice

(c) document images

(d) other images

Fig. 1 The samples of finance images **a** other invoice **b** document images **c** other images

can reduce the input dimension and reduce the computation resources; finally, using the prediction from the four CNN models to generate the prediction result.

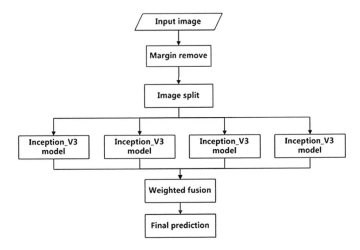

Fig. 2 Flow graph of our classification method

Fig. 3 The example of contains many objects

2.1 Preprocessing

The resolution of finance images is typically higher than 1500 * 1500, which is too large for CNN training with current availability of computing resources. Besides, the margins of finance images (as Fig. 1 described) also contain many useless features. Large input dimensionality and many margins not only cost too many computation resources but also lead to a high probability of overfitting.

Image margin removing. As the Fig. 3 described, we can find that the images always content many margins which can cause the overfitting problem in training stage. Through an analysis of finance images, we find out that the margins of image can be removed by image processing method. In this paper, in order to filter the margins of images, we propose a simple but effective strategy to obtain the refined image.

First, we transform the color image into grayscale image according to the following function:

$$P_{\text{Gray}} = R * 0.229 + G * 0.587 + B * 0.114 \tag{1}$$

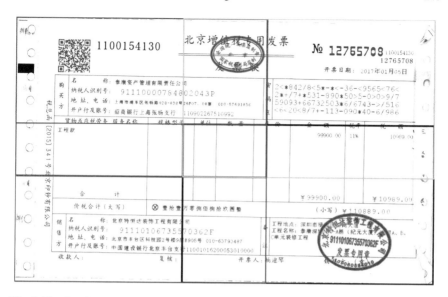

Fig. 4 The four image patches segment from the true image, where the red rectangle stand for $P_1(x, y)$, the green rectangle stand for $P_2(x, y)$, the blue rectangle stand for $P_3(x, y)$, the black rectangle stand for $P_4(x, y)$ respectively

Then we use the Otsu's algorithm [14] to obtain the binary image in which black regions are the background and white regions are the target regions. As shown in Fig. 3, the image usually contains many targets, so we select the Minimum Bounding Rectangle of these object regions as the refined image. This method can filter almost all margins of finance images with a small amount of computing resources.

Dividing the refined image into small patches. The resolution of finance images is typically higher than 1500 * 1500, which is too large for neural network. To overcome such a large size problem, we propose to use the sub-images as the input of neural network instead of the full image. Specifically, we divide the refined image into four image patches according to the following strategy. In order to avoid the feature loss when segment, we make the segment contains some overlap regions (as describe as Fig. 4). Suppose the size of true image $I(M, N)$ is $M \times N$, and the upper left corner is the origin of the coordinates. Thus we use the follow formulations to obtain four image patches

$$P_1 = I(0 : 0.6M, 0 : 0.6N) \tag{2}$$

$$P_2 = I(0.4M : M, 0 : 0.6N) \tag{3}$$

$$P_3 = I(0 : 0.6M, 0.4N : N) \tag{4}$$

$$P_4 = I(0.4M : M, 0.4N : N), \tag{5}$$

where P_i (i=1, ..., 4) is the image patches, and the first parameter of I stand for x coordinate, the second parameter of I stand for y coordinate respectively.

In this step, we can divide one image into four different image patches. On the one hand, this strategy can provide more training examples for feature learning, on the other hand, we can through this method to reduce the impact of reducing the input dimension.

2.2   The Neural Network Apply

CNN has been proved very successful in solving the image classification problem. Research works based on CNN have made significantly improvement for many image databases, including the MNIST database, the NORB database and the CIFAR10 dataset. CNN is good at learning the local and global structures from image data.

In this paper, we present a general method for finance image classification using inception_v3 neural network [15]. Inception-v3 is designed for the ImageNet Large Visual Recognition Challenge using the data from 2012. This structure can lead to high performance vision networks that have a relatively modest computation cost compared to simpler, more monolithic architectures [16]. For the image patches obtained from Sect. 2.2, their sizes are usually different, which need to be resized to a unified size. We set the unified size to 299 * 299. Based on Inception-v3 model of TensorFlow platform, transfer learning technology was used to retrain our datasets, which can greatly improve the accuracy of finance image classification.

2.3   The Weighted Fusion and Prediction

For the probability of four image patches according to the above-mentioned steps, we can get four prediction results as output. How to detect the final categories of finance images is very complex. In our experiment, we set the same weights for different location of the images. The following equation is used to calculate the final prediction:

$$\mathrm{Pre}_y = 0.25 * P_1 + 0.25 * P_2 + 0.25 * P_3 + 0.25 * P_4, \tag{6}$$

where Pre_y is the final prediction of the input images, $P_{i(i=1...4)}$ is the prediction of four sub-images.

Table 1 The result of without filtering the margins [15]

	VAT invoice	Other invoice	Document images	Other images
VAT invoice	287	6	4	3
Other invoice	3	263	21	13
Document images	0	0	274	26
Other images	2	3	20	275
Accuracy (%)	95.7	87.7	91.3	91.7

Table 2 The result of VGG19 networks [16]

	VAT invoice	Other invoice	Document images	Other images
VAT invoice	267	16	8	9
Other invoice	22	237	11	16
Document images	7	22	245	26
Other images	0	12	15	273
Accuracy (%)	89	79	81.7	91

Table 3 The result the proposed method

	VAT invoice	Other invoice	Document images	Other images
VAT invoice	276	10	10	4
Other invoice	2	273	14	11
Document images	0	0	280	20
Other images	3	10	12	275
Accuracy (%)	92	91	93.3	91.7

3 Experiment

In this paper, in order to prove our method, we used different networks (such as VGG19 [16]) to train the classification models without filtering the image margins. We used 3000 images for each category, and total 12,000 images for training. Besides, in order to evaluate our method, we also select 300 images for each category and total 1200 images from our image system for testing. This test images is total different from the train sets (Tables 1, 2 and 3).

First, we took the preprocessing steps as shown in Sect. 2.1 without filtering the margins of the images, and thus the results are as follows:

Second, we used VGG19 network structure to train the models, and the result are as follows:

Finally, we used the proposed method to train the models, and thus the result are as follows:

We can know that the proposed method has the best results, and it can well deal with the finance images classification problem.

4 Conclusion

The proposed method has the following noticeable merits: first, we use a simple method to get the true image, and use a fixed model to segment four sub-image. In our experiment, we found that this step can avoid the large-scale problem for training the models. Our method can remove some margins of image which results in an overfitting problem. Second, we use the transfer learning technology to retrain our datasets, which can greatly improve the accuracy of finance image classification. Third, the proposed classification method can efficiently and precisely predict the category of an input image. The experimental results show that the proposed method performs very well on our databases comparing with other methods.

References

1. Milgram, J., Sabourin, R., Cheriet, M.: Two-stage classification system combining model-based and discriminative approaches. In: The 17th International Conference on Pattern Recognition, pp. 152–155 (2004)
2. Farid, M., Lorenzo, B.: Classification of hyperspectral remote sensing images with support vector machines. IEEE Trans. Geosci. Remote Sens. **42**(8), 1778–1790 (2004)
3. Xu, Y., Zhang, B., Zhong, Z.: Multiple representations and sparse representation for image classification. Pattern Recogn. Lett. **68**, 9–14 (2015)
4. Guo, S., Huang, W., Qiao, Y.: Improving scale invariant feature transform with local color contrastive descriptor for image classification. J. Electron. Imaging **26**(1), 013015 (2017)
5. Guo, Z.H., Zhang, L., Zhang, D.: A completed modeling of local binary pattern operator for texture classification. IEEE Trans. Image Process. **19**(6), 1657–1663 (2010)
6. Zhou, L., Wang, L., Liu, L., Ogunbona, P., Shen, D.: Support vector machines for neuroimage analysis: interpretation from discrimination. In: Support Vector Machines Applications, pp. 191–220. Springer, Cham (2014)
7. Sun, G., Chen, T., Su, Y., et al.: Internet traffic classification based on incremental support vector machines. Mobile Netw. Appl. **23**(120), 1–8 (2018)
8. Xu, Y., Zhu, Q., Fan, Z., Zhang, D., Mi, J., Lai, Z.: Using the idea of the sparse representation to perform coarse-to-fine face recognition. Inf. Sci. **238**(20), 138–148 (2013)
9. Rawat, W., Wang, Z.: Deep convolutional neural networks for image classification: a comprehensive review. Neural Comput. **29**(9), 1 (2017)
10. Williams, T., Li, R.: An ensemble of convolutional neural networks using wavelets for image classification. J. Softw. Eng. Appl. **11**(2), 69–88 (2018)
11. Sharma, A., Liu, X., Yang, X., et al.: A patch-based convolutional neural network for remote sensing image classification. Neural Netw. **95**, 19–28 (2017)
12. Steinkrau, D., Simard, P.Y., Buck, I.: Using GPUs for machine learning algorithms. In: Proceedings of the 8th International Conference on Document Analysis and Recognition, Washington, pp. 1115–1119 (2005)

13. Hinton, G.E., Osindero, S., Teh, Y.: A fast learning algorithm for deep belief nets. Neural Comput. **18**(7), 1527–1554 (2006)
14. Bhargava, D.N., Kumawat, A., Bhargava, D.R.: Threshold and binarization for document image analysis using otsu's Algorithm. Int. J. Comput. Trends Technol. **17**(5), 272–275 (2014)
15. Xia, X., Xu, C., Nan, B.: Inception-v3 for flower classification. In: 2017 2nd International Conference on Image, Vision and Computing (ICIVC), Chengdu, pp. 783–787 (2017)
16. Simon, M., Rodner, E., Denzler, J.: ImageNet pre-trained models with batch normalization, pp. 1–4 (2016)

Rough Possibilistic Clustering for Fabric Image Segmentation

Jie Zhou, Can Gao and Jia Yin

Abstract Fabric image segmentation is very important for fabric designing, fabric manufacturing and textile printing and dyeing. The results of fabric image segmentation are inevitably influenced by some noisy factors, such as fibers, dust, stains, holes, neps, and other fabric defects. In this paper, a novel fabric image segmentation method based on rough possibilistic clustering is presented. Specifically, all pixels are partitioned into three approximation regions with respect to a fixed cluster, i.e., the core, boundary, and exclusive regions of this cluster. The corresponding prototype calculation is only related to the core and boundary regions, rather than all the image pixels. Thus, the obtained prototypes cannot be distorted by the pixels belonging to other colors or regions. In addition, since the typicality values are involved, the proposed method is robust for dealing with noisy environments. The improved performance of the proposed method is illustrated by some real fabric images.

Keywords Rough possibilistic clustering · Approximation region · Uncertainty
Fabric image segmentation

1 Introduction

Fabric image segmentation is very important for fabric designing, fabric manufacturing and textile printing and dyeing. The results of fabric image segmentation are inevitably influenced by some noisy factors, such as fibers, dust, stains, holes, neps and other fabric defects [1]. How to precisely segment the given fabric images with machine learning technologies will improve the automation level of textile industries.

J. Zhou (✉) · C. Gao
College of Computer Science and Software Engineering,
Shenzhen University, Shenzhen 518060, Guangdong, China
e-mail: jie_jpu@163.com

J. Yin
Kuang Yaming Honors School, Nanjing University, Nanjing 210023,
Jiangsu, China

© Springer Nature Switzerland AG 2019
W. K. Wong (ed.), *Artificial Intelligence on Fashion and Textiles*,
Advances in Intelligent Systems and Computing 849,
https://doi.org/10.1007/978-3-319-99695-0_30

Clustering [2] focuses on partitioning a data set with unlabeled patterns into some subgroups, and clustering methods are often used for image segmentation [3]. However, when it comes to the fabric images, the noisy factors need to be addressed. Possibilistic approach to clustering, especially possibilistic C-means (PCM) [4], was proposed to address the noisy environments. Since a faraway pattern, i.e., noisy data or outliers, would belong to the clusters with very small possibilistic memberships (typicality values), PCM is more robust than fuzzy C-means [5] as handling data in a noisy environment. However, if the initializations of the iteration implementations of PCM are not good, the performance of PCM tends to coincidental clusters. In order to overcome this limitation, Pal et al. [6] proposed a possibilistic-fuzzy C-means model (PFCM), which integrates both typicality values and fuzzy membership degrees in clustering processes. Xenaki et al. [7] introduced sparse possibilistic C-means that can deal with closely located clusters.

Rough set theory [8] provides a new methodology for analyzing data involving uncertain, imprecise or incomplete information. Lingras et al. [9] proposed a rough C-means (RCM) clustering method. Each cluster is described not only by a prototype but also with a pair of lower and upper approximations. Meanwhile, the boundary region is defined as the difference between the lower and upper approximations. The uncertainty and vagueness arising in the boundary region of each cluster can be captured well in RCM. Mitra et al. [10] developed rough-fuzzy possibilistic C-means (RFCM) in which the fuzzy membership degrees are used instead of absolute distances for approximation region partitions. However, how to select the partition threshold is rarely discussed in these studies. Different partition thresholds will result in different approximation regions, and thereafter the prototype calculations will be discrepant.

This study will focus on rough possibilistic clustering for fabric image segmentations which involves two phases. In the first phase, all pixels are partitioned into three approximation regions with respect to a fixed cluster. In the second phase, the classification labels of pixels that partitioned into the absolute exclusion regions over all clusters will be modified with the labels of its nine neighbors. In this way, the misclassification caused by the noisy pixels can be reduced.

The rest of paper is organized as follows. In Sect. 2, a rough possibilistic clustering method based on shadowed sets in presented. In Sect. 3, fabric image segmentation based on the proposed method is introduced. Comparative experimental results are shown in Sect. 4. Some conclusions are drawn in Sect. 5.

2   Rough Possibilistic Clustering Based on Shadowed Sets

In this section, a rough possibilistic clustering method based on shadowed sets is presented. More details about PCM and rough sets can be found in [4, 8].

According to the shadowed set optimization mechanism [11], the obtained possibilistic membership degrees of all patterns to a fixed cluster can be divided into three levels, i.e., high enough, low enough, and the shadows, based on which three approximation regions can be formed with respect to this cluster. By integrating the notions of shadowed sets and PCM, a rough possibilistic C-means based on shadowed sets can be presented in Algorithm 1.

From the Steps 3 to 5 in Algorithm 1, the threshold of each cluster is not user-defined beforehand. It can be adjusted automatically in the clustering processes and can be optimized for each cluster independently.

According to Eq. (1), after partitioning all patterns into approximation regions for all clusters, the patterns $\{\mathbf{x}_j\}$ will belong to the three situations as follows:

S1: belonging to the core region of at least one cluster, i.e., $\exists G_i$, such that $\mathbf{x}_j \in \underline{R}G_i$;
S2: not belonging to the core regions of any clusters, but belonging to the boundary region of at least one cluster, i.e., $\forall G_i \ (i = 1, 2, \ldots, C)$, it has $\mathbf{x}_j \notin \underline{R}G_i$, but $\exists G_p$, such that $\mathbf{x}_j \in R_b G_p$;
S3: not belonging to the core regions and boundary regions of any clusters, i.e., $\forall G_i (i = 1, 2, \ldots, C)$, it has $\mathbf{x}_j \notin \underline{R}G_i$ and $\mathbf{x}_j \notin R_b G_i$.

If a pattern belongs to the core region of one cluster as well belongs to the boundary regions of the other clusters, this situation belongs to **S1**. If a pattern belongs to **S3**, it can be considered as a noisy data or an outlier.

Algorithm 1 Rough possibilistic clustering based on shadowed sets (sRPCM)

Step 1: Initialization of the prototypes $\mathbf{v}_i (i = 1, 2, \ldots, C)$
Step 2: Compute scale parameters $\{\gamma_i\}$;
Step 3: Compute possibilistic membership values μ_{ij};
Step 4: Based on shadowed sets, compute optimal threshold α_i for each cluster according to possibilistic membership values;
Step 5: According to α_i, determine the core and the boundary regions of each cluster G_i:

$$\underline{R}G_i = \left\{\mathbf{x}_j | \mu_{ij} \geq \max_j(\mu_{ij}) - \alpha_i\right\} \quad R_b G_i = \left\{\mathbf{x}_j | \alpha_i < \mu_{ij} < \max_j(\mu_{ij}) - \alpha_i\right\};$$
(1)

Step 6: Compute the prototypes $\mathbf{v}_i (i = 1, 2, \ldots, C)$ as that in RFCM, but with possibilistic membership values;
Step 7: Repeat Steps 2 to 6 until convergence is achieved.

3 Fabric Image Segmentation Based on SRPCM

Generally, fabrics often have some defects, such as adhesions, stains, holes, neps and other fabric defects. Inevitably, fabric images will be with some noisy pixels which can affect the results of segmentation. According to the rough possibilistic clustering based on shadowed sets, the noisy data or outliers can be partitioned into the absolute

Algorithm 2 Fabric image segmentation based on sRPCM

Step 1: Obtain the prototypes, possibilistic memberships μ_{ij} and approximation region partitions of each cluster based on sRPCM;

Step 2: $\forall \mathbf{x}_j$, assign its classification labels as $\text{Lable}(\mathbf{x}_j) = q$, if $\mu_{qj} = \max\limits_i(\mu_{ij})$;

Step 3: $\forall \mathbf{x}_j$, if it satisfies **S3**, denotes $Nlable(\mathbf{x}_j) = 1$, otherwise, $N\text{lable}(\mathbf{x}_j) = 0$;

Step 4: $\forall \mathbf{x}_j$, if $N\text{lable}(\mathbf{x}_j) = 1$, then update its classification label as follows: Compute the frequency of labels of its nine-neighborhood pixels with $N\text{lable}(\mathbf{x}_j) = 0$, and select the value with the maximum frequency as the final label of $\mathbf{x}_j$;

Step 5: If a pattern with $N\text{lable}(\mathbf{x}_j) = 1$, and all of its nine-neighborhood pixels are also with $N\text{lable}(\mathbf{x}_j) = 1$, then compute the distances between this pattern and all prototypes, and assign its label as $\text{Lable}(\mathbf{x}_j) = q$, if $d_{qj} = \min\limits_i(d_{ij})$.

exclusion region over all clusters, i.e., they can be divided into **S3** to the greatest extent. In this case, fabric image segmentation based on sRPCM can be described in Algorithm II which can be separated into two phases. The first phase includes Steps 1 and 2, in which the classification label of each pattern is determined according to its possibilistic memberships over all clusters. The second phase includes Steps 3 to 5, in which the classification labels of the patterns belonging to **S3** are modified based on the properties of its nigh-neighbors. In this way, the final classification labels of noisy pixels can be revised.

4 Experiments

In this section, two fabric images are involved for experiments. The fuzzification coefficients for fuzzy and possibilistic clustering methods are set as 2 and 1.5, respectively. The weighted value is set as $w_l = 0.95$ and kept as a constant. The maximum iteration number is set as 100 and the convergence condition satisfies $\|\mathbf{v}_i(t+1) - \mathbf{v}_i(t)\| < \varepsilon$ where t is an iterative step and ε is set as 0.001.

4.1 Fabric Image I

The original fabric image I and its ground-truth can be found in Fig. 1a, b respectively.

Since the influence caused by fibers, the original fabric image I seems to be blurring to some extent. The results of segmentations generated by four clustering methods, i.e., FCM, PCM, PFCM and the proposed method are shown in Fig. 1c, f. The corresponding classification accuracies are compared in Fig. 2a. It can be found that the proposed method achieves the best classification accuracy for the original image.

To simulate the ground-truth image under different noisy environments, some salt-and-pepper noises are involved. The classification accuracies obtained by four

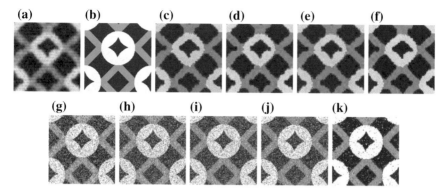

Fig. 1 The segmentation results for fabric image I. **a** original image; **b** ground-truth; **c** FCM; **d** PCM; **e** PFCM; **f** sRPCM; **g** noisy image (30%); **h** FCM for noisy image; **i** PCM for noisy image; **j** FPCM for noisy image; **k** sRPCM for noisy image

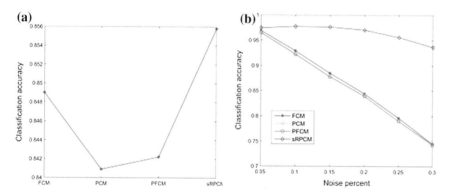

Fig. 2 The classification accuracy for fabric image I. **a** the classification accuracy for original image; **b** the classification accuracies for noisy images

clustering methods can be found in Fig. 2b. As increasing the percent of noisy data, the performances of FCM, PCM and FPCM are decreasing fast. Though the performance of the proposed method sRPCM is also decreasing, the degree of change is smaller than other methods. For the noisy image with 30% noisy data, as shown in Fig. 1g, the results obtained by four methods can be shown in Fig. 1h, k. Obviously, most noisy data are filtered out in the result obtained by sRPCM.

4.2 Fabric Image II

The original fabric image II, which is a scanned image, and its ground-truth can be found in Fig. 3a, b, respectively.

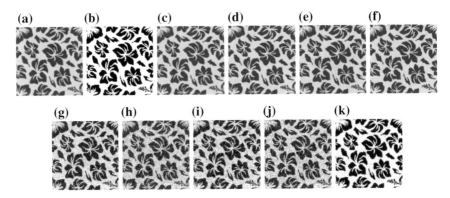

Fig. 3 The segmentation results for fabric image II. **a** original image; **b** ground-truth; **c** FCM; **d** PCM; **e** PFCM; **f** sRPCM; **g** noisy image (30%); **h** FCM for noisy image; **i** PCM for noisy image; **j** FPCM for noisy image; **k** sRPCM for noisy image

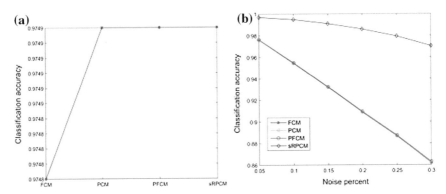

Fig. 4 The classification accuracy for fabric image II. **a** The classification accuracy for original image; **b** the classification accuracies for noisy images

The segmentation results generated by four clustering methods are shown in Fig. 3c, f. The corresponding classification accuracies are compared in Fig. 4a. Since only two clusters need to be grouped, i.e., black and white pixels, the four methods achieve good performances. However, as increasing the salt-and-pepper noises to the ground-truth image, the performances obtained by four methods are changed, as shown in Fig. 4b. The performances of FCM, PCM, and PFCM are decreasing drastically. The decline generated by the proposed method is relatively smaller compared with other methods. For the noisy image with 30% noisy data, as shown in Fig. 3g, the results obtained by four methods are shown in Fig. 3h, k. Visually, the result obtained by sRPCM is closer to the ground-truth image than another methods.

5 Conclusions

In this study, a rough possibilistic clustering method based on shadowed sets is presented for fabric image segmentation, in which all the pixels are divided to three situations, i.e., belonging to the core region of at least one cluster, belonging to the boundary region of at least one cluster, not belonging to the core and boundary regions of any clusters. In this case, the noisy pixels can be well detected, and their classification labels are modified according to the properties of their neighbors. The experiments on the real fabric images illustrate the validity of the proposed method. The fuzzifier parameter is pivotal for the fuzzy and possibilistic clustering methods, the uncertainties caused by this parameter will be analyzed in our future works.

Acknowledgements The work is supported by Postdoctoral Science Foundation of China (No. 2017M612736, 2017T100645), Guangdong Natural Science Foundation (No. 2018A030310450, 2018A030310451).

References

1. Kumar, A.: Computer-vision-based fabric defect detection: a survey. IEEE Trans. Ind. Electron. **55**(1), 348–363 (2008)
2. Aggarwal, C.C., Reddy, C.K.: Data clustering: algorithms and applications. Chap man and Hall/CRC Press, Boca Raton, Florida (2013)
3. Dhanachandra, N., Manglem, K., Chanu, Y.J.: Image segmentation using K-means clustering algorithm and subtractive clustering algorithm. Procedia Comput. Sci. **54**, 764–771 (2015)
4. Krishnapuram, R., Keller, J.M.: A possibilistic approach to clustering. IEEE Trans. Fuzzy Syst. **1**, 98–110 (1993)
5. Bezdek, J.C.: Pattern recognition with fuzzy objective function algorithms. Kluwer Academic Publishers, Norwell, MA, USA (1981)
6. Pal, N.R., Pal, K., Keller, J.M., Bezdek, J.C.: A Possibilistic fuzzy c-means clustering algorithm. IEEE Trans. Fuzzy Syst. **13**(4), 517–530 (2005)
7. Koutroumbas, K.D., Xenaki, S.D., Rontogiannis, A.A.: On the convergence of the sparse possibilistic C-means algorithm. IEEE Trans. Fuzzy Syst. **26**(1), 324–337 (2018)
8. Pawlak, Z.: Rough sets. Int. J. Comput. Inf. Sci. **11**(5), 314–356 (1982)
9. Lingras, P., West, C.: Interval set clustering of web users with rough k-means. J. Intell. Inf. Syst. **23**(1), 5–16 (2004)
10. Mitra, S., Banka, H., Pedrycz, W.: Rough-fuzzy collaborative clustering. IEEE Trans. Syst. Man. Cybern. B **36**(4), 795–805 (2006)
11. Pedrycz, W.: From fuzzy sets to shadowed sets: interpretation and computing. Int. J. Intell. Syst. **24**, 48–61 (2009)

Fashion Meets AI Technology

Xingxing Zou, Wai Keung Wong and Dongmei Mo

Abstract With the development of Artificial Intelligence (AI) technology, extensive research efforts have been devoted to the cross-disciplinary area of fashion and AI. This paper reviews previous research studies of AI on fashion aspect. Fundamental image processing technologies will be introduced first and followed by apparel recognition, and fashion aesthetic understanding. We finally propose a framework about the research direction of AI research on fashion as a reference for inspiring fashion and AI-related researchers.

Keywords Fashion · AI · Aesthetic understanding

1 Introduction

The development of technology can always reshape the industry and change people's behavior. For example, with the advancement of image retrieval technology, people can easily find the clothes they like and purchase through the online e-commerce platform (street-to-shop [1]). To cater for more people concerning very much on beauty, more and more researchers are beginning to shift their research direction from apparel recognition to fashion understanding.

However, fashion understanding is actually a very complicated issue because its essence is to study the aesthetics of apparel. Thus, the first question that needs to be answered is, what is beauty? How can it be considered as beautiful? This is a philosophical issue. Since the eighteenth century, philosophers have been discussing this issue. Different genres based on their different positions, such as the standpoint of materialism and idealism, make their definition of aesthetics. More consensus is

X. Zou · W. K. Wong (✉)
Institute of Textiles and Clothing, The Hong Kong Polytechnic University, Hong Kong, China
e-mail: calvin.wong@polyu.edu.hk

D. Mo
College of Computer Science and Software Engineering,
Shenzhen University, Shenzhen, China

© Springer Nature Switzerland AG 2019
W. K. Wong (ed.), *Artificial Intelligence on Fashion and Textiles*,
Advances in Intelligent Systems and Computing 849,
https://doi.org/10.1007/978-3-319-99695-0_31

that beauty is a feeling that makes people feel happy. Beauty or deformity in an object results from its nature or structure. To perceive the beauty, therefore, we must perceive the nature or structure from which it results. In this, the internal sense differs from the external. Our external senses may discover qualities which do not depend upon any antecedent perception. But it is impossible to perceive the beauty of an object, without perceiving the object, or at least conceiving it [2].

Artists usually use their artwork, no matter it is a song, a painting, or an apparel, to express themselves and then communicate with others. Perception is the most important key word in aesthetic. However, for a computer, at least, we do not want it to obtain the perception ability. Fortunately, perceiving the nature or structure of an object is one thing and perceiving its beauty is another [3]. Thus, if we just consider a basic level, the fashion understanding tasks of the computer still can be solved. In this paper, we review previous research studies of AI on fashion aspect. Fundamental image processing technologies will be introduced first and followed by apparel recognition, and fashion aesthetic understanding. We finally propose a framework about the research direction of AI research on fashion.

2   Previous Works

Over past decades, many AI researchers devoted efforts in the fashion area. In this section, we review and summarize their works in the past decades.

2.1   Fundamental Technologies

In 2006, Chen and Xu [4] proposed a template for cloth modeling and sketching. The design components such as collar, shoulder, and sleeves were used to present an apparel together. Through the edge map, bounding detection, and skin color detection, they could obtain the sketch of apparel. Yang and Yu [5] proposed the clothing representation combining color histograms and three texture descriptors, including Histogram of Oriented Gradients (HOG), Bag-of-Words (BoW), and Discrete Cosine Transform (DCT), to realize real-time clothing recognition in surveillance videos. However, only front view of images could be processed.

In 2015, Liu et al. [6] proposed a novel semi-supervised learning strategy to realize fashion parsing in diverse poses. Additionally, most researches were devoted to fashion parsing in photographs. Wang and Ai et al. [7] focused on clothing segmentation for highly occluded group images. In 2012, Yamaguchi et al. [8] proposed a method with Conditional Random Field (CRF) model for fashion images with standing model in the front view. A year later, she with her team member [9] further improved their method that combines parsing from pretrained global clothing models and local clothing models learned from retrieved examples, and transferred parse masks (paper doll item transfer) from retrieved examples.

Several reported studies further improved the performance of the fashion parsing. Dong et al. [10] proposed a deformable mixture parsing model. Yang et al. [11] used segmentation and labeling to joint parsing the apparel. In addition, Liu et al. [12] addressed the problem of traditional fully supervised algorithms with weak color-category labels. Simo-Serra et al. [13] proposed an approach based on CRF model for clothing parsing.

In 2015, Yamaguchi introduced a data-driven approach to clothing parsing [14] based on her former research [9]. They collected a large, complex, real-world collection of outfit from the website with 339,797 images and used extensive experiments to demonstrate their advances by comparison with the previous method such as localization (clothing parsing given weak supervision in the form of tags) or detection (general clothing parsing). In 2017, to satisfy the new requirement of semantic understanding, Kota Yamaguchi with her team member further introduced the extended fully convolutional neural networks [15] to provide higher level knowledge on clothing semantics and contextual cues to disambiguate fine-grained categories for clothing parsing. Additionally, Neuberger et al. [16] introduce a closed-form expression to model the label uncertainty induced by subsampling.

2.2 Fine-Grained Apparel Recognition

In 2011, Li et al. [17] used sparse representation to realize visual element analysis. The visual fashion elements, especially the color of apparel, were used to retrieve similar apparel. In addition, color matching and attribute learning based apparel retrieval were also proposed by Wang and Zhang [18]. The low-level features like color were mainly taken into consideration, and the clothes attributes including the type of clothes, sleeves, and patterns were used as the reference.

Then, street-to-shop task was presented by Liu et al. [1] in 2012. The cross-scenario image retrieval task was tackled via parts alignment and apparel attributes. Further, Fu et al. [19] detected human parts as the semantic and encoded it into the vocabulary tree under the Bag-of-visual-Word (BoW) framework and retrieval apparel with semantic-preserving visual phrases. Chen et al. [20] proposed bundled features which consist of the point features (SIFT). At the same year, Chen et al. [21] employed the human pose estimation by Kinect sensors to improve the performance of clothing recognition. With the hot research focus on sparse coding, Huang et al. [22] proposed sparse coding to clothing image retrieval.

Additionally, Bossard et al. [23] adopted Support Vector Machine (SVM) to conduct the multiple classification learning for apparel classification. In 2013, Wei and Catherine [24] aimed at fine-grained recognition in women's fashion coat using learning model. Then, Wang et al. [25] built two types of models including a color-based BoW model and general reranking models of clothing attributes. Kalantidis et al. [26] and Kiapour et al. [27] focused on the daily photos which are connecting to online products. Apparels in daily photos could be recognized and searched on the commercial websites.

To further improve the performance of image retrieval, Huang et al. [28] proposed a dual attribute-aware ranking network for feature learning. Mizuochi et al. [29] used local regions of multiple images to search similar clothing. Users can retrieve their desired clothes which are globally similar to an image and partially similar to another image. For fine-grained clothing attributes recognition, Chen et al. [30] adopted Region CNN (RCNN) to localize human bodies and presented a double-path deep domain network. The framework of [30] is shown in Fig. 1.

To shorten the searching time of clothing retrieval, Lin et al. [31] developed a hierarchical deep search framework via deep learning of binary codes. Additionally, Veit et al. [32] proposed a siamese convolutional neural network for the style of apparel matching. Yamaguchi et al. [33] used eight bounding boxes with 58 attributes to realize clothing recognition for mix and match. Unlike the bounding box used in [33], Sun et al. [34] divided the clothing images into different parts based on the structure of human body. Simo-Serra and Ishikawa [35] proposed a feature learning approach using weak data for small-scale training. Vittayakon et al. [36] also introduced an unsupervised approach with weak annotations to automatically discover the clothing attributes. Saliency detection was used in their framework. Different from the previous methods, Liu et al. [37] first proposed to use the key points of the apparel for fashion landmark detection. They further published fashion dataset [38] which is the largest scale fashion dataset until now. In addition, Sengupta et al. [39] proposed a scheme of feature encoding. Chen and Luo [40] introduced a novel system which can obtain useful insights on clothing consumption trends and profitable clothing features. Laenen et al. [41] developed a neural network which learns intermodal representations for fashion attributes to be utilized in a cross-modal search tool.

2.3   Fashion Understanding

Fashion aesthetic understanding is not only to recognize the attributes of an apparel but also understand the meanings behind those attributes. In 2011, Iwata et al. [42]

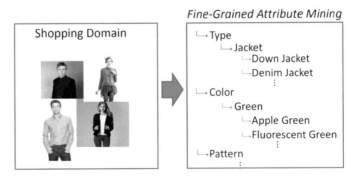

Fig. 1 Overview of the proposed approach in [30] which used fine-grained attributes

proposed a framework of fashion recommendation system which was based on region detection and visual feature similarities matching. Liu et al. [43, 44] introduced a concept of "magic mirror" for fashion recommendation including hairstyle, makeup, and mix and match.

Yamaguchi et al. [45] used the fashion image from the website to predict fashion trend based on the social comment. Jagadeesh et al. [46] proposed a framework, which is shown in Fig. 2, of large-scale visual recommendations from street fashion images. Vittayakon et al. [47] presented an approach to studying fashion both on the runway and the real-world settings. Kiapour et al. [48] built a style dataset which selected five different styles including hipster, bohemian, pinup, preppy, and goth.

Simo-Serra et al. [49] used online users' photos to train a model to give a feedback on the personal mix and match suggestions to every user. Vaccaro et al. [50] presented fashion applications that allow users to express their desires in natural language and the proposed system was like a real personal stylist which can suggest recommendations and customized items to the users.

Al-Halah et al. [51] proposed to forecast visual style trends before they occur. The first approach was to predict the future popularity of styles discovered from fashion images in an unsupervised manner. Using these styles as a basis, a forecasting model could be trained to represent their trends over time. Kang et al. [52] proposed a new approach, which is shown in Fig. 3, to generate new images that are most consistent with their personal taste.

Qian et al. [53] introduced a recommendation system which is imbedded with the knowledge from experts (images from fashion blogs). Lin et al. [54] also developed a fashion recommendation system. They studied the stylists on the social network and use topic modeling to learn the underlying topics of brands, items, and colors. Han et al. [55] focused on three tasks: recommending a fashion item that etches the style of an existing set; generating an outfit based on user's text/image inputs; and

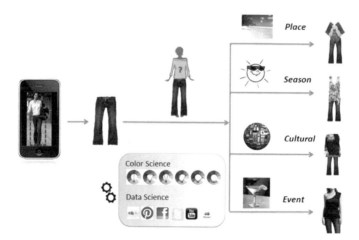

Fig. 2 Overview of the proposed approach in [46]

Fig. 3 Personal recommendation and design framework in [52]

predicting the compatibility of an outfit. He and McAuley [56] build novel models for the one-class collaborative filtering setting for estimating users' fashion-aware personalized ranking functions based on their previous feedback.

Garg et al. [57] formulated a mechanism that grades the look of a product that captured visual aesthetics for obtaining the analysis of sales potential. Lee et al. [58] used the data from Twitter to identify whether a Twitter account is fashion related. Arora et al. [59] proposed a creation of fashion sensibilities which can be thought of as commonly occurring fashion tastes. Mohammed Abdulla and Borar [60] proposed a size recommendation system for fashion e-commerce.

Lorbert et al. [61] introduced a bidirectional GAN framework to reconstruct style images. Date [62] constructed an approach to personalize and generate new customer clothes based on a user's preference. The overview of its system architecture is shown in Fig. 4. Also, Zhu et al. [63] presented an approach to generate new clothing on a wearer through generative adversarial learning.

3 Datasets and Applications

In this section, we first summarize the typical fashion datasets, including its size, source, and use, and then report some representative AI applications on fashion industry.

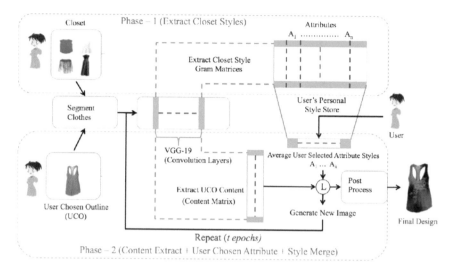

Fig. 4 Overall system architecture in [62]

3.1 Fashion Datasets

Cross-scenario Clothing Retrieval: (1) **OS/DP dataset** [1] was a large online shopping dataset and a daily photo dataset, both of which are thoroughly labeled with 15 clothing attributes. (2) **Exact Street2Shop dataset** [27] provided a total of 39,479 clothing item matches between street and shop photos with 11 categories. (3) **DARN dataset** [28] consists of 450,000 online shopping images and 90,000 exact offline counterpart images with 179 attributes. (4) **DDA dataset** [30] was built with 341,021 images with 67 attributes.

　　Attributes Recognition: (1) [23] defined 15 clothing classes and introduced a benchmark dataset consisting of over 80,000 images. (2) [64] presented a dataset containing a mixture of 4810 images composing of general images and fashion-related images. (3) **Chictopia dataset** [33] contains 268,124 images with 58 attributes. (4) **What-to-Wear dataset** [43] contains 24,417 labeled fashion images with seven multi-value clothing attributes and 10 occasion categories. (5) **Style dataset** [48] collected 1893 images with five different styles. (6) **Fashion 144K dataset** [49] collected 144,169 user posts containing diverse image, textual, and meta information. (7) **Runway dataset** [47] collected 348,598 images from style.com. (8) **Fashion-MNIST** [65] comprising of gray scale images of 70,000 fashion products from 10 categories. 9) **Sketch Me That Shoe** [66] introduces a new database of 1432 sketch–photo pairs from two categories with 32,000 fine-grained triplet ranking annotations. (10) **UT-ZAP50K** [67, 68] consists of 50,025 catalog images which were divided into four major categories: shoes, sandals, slippers, and boots.

　　Parsing Clothing: (1) **Fashionista dataset** [8] consists of 158,235 photographs collected from Chictopia.com. Also, it contains 685 fully parsed images. (2) **Paper**

Fig. 5 Instances in [71, 72], from left to right the body shape attributes shown are apple, column, hourglass, and pear

Doll dataset [9] collected over 1 million pictures from chictopia.com with associated metadata tags. (3) **Clothing Co-parsing dataset** [11] consists of 2098 high-resolution fashion photos with huge human/clothing variations. (4) **Colorful-Fashion dataset** [12] contains 2682 images with pixel-level color-category labels. (5) **Fashion Icon dataset** [6] contains 1500 unlabeled video downloaded from youtube.com.

Image Retrieval: (1) **WFC dataset** [24] obtained the 12 coats/jackets subcategories from eBay with 2092 images. (2) **Fashion 10000 dataset** [69] contains 32,000 images with 262 different fashion categories. (3) **DeepFashion dataset** [38] contains over 800,000 images, which were richly annotated with massive attributes. (4) **Deep-Fashion Alignment dataset** [37] contains 120,000 images with eight landmarks.

Occupation Recognition: (1) [70] presented **the occupation database** with 7000 images of 14 different occupations.

Aesthetic Evaluation: (1) [71, 72] collected 1064 images for 120 configurations where a configuration represents an image of a person of a body shape with specific top and bottom clothing categories. The instance in this dataset is shown in Fig. 5. (2) StreetStyle dataset [73] explores world-wide clothing styles from millions of photos.

3.2 Industry Applications

The main reason for AI technology becoming the increasingly hot topic is its surprising performance in real applications which are highlighted in this section.

IBM: Use IBM Research AI tools to analyze real-time images and fashion industry trends.

Frilly: Allow users to fully customize their clothing on their website.

Trendage: Generate more than 10 million style recommendations a month for apparel, accessories, and footwear retailers.

Sephora Visual Artist: Allow potential customers to "try on" cosmetic products including lipsticks, eye shadows, and highlight palettes.

Thread: Pair customers with a stylist and create tailored recommendations based on the customer's stylistic preferences.

Amazon Echo Look: Help users select the right outfit.

Shiseido: Allow virtual makeup try on, tutorials, color matching, personalized recommendations, makeup removal, face tracking, and skin tone detection.

ASOS: Allow the user's smartphone camera to take a picture of a product by identifying the shape, color, and pattern of the object.

Google: Publishes a fashion trend report in 2016.

4 Discussions

Based on the review in the above sections, for the feature vector matching to attributes learning, we can see that the current AI technology can well match up the similar apparel images. Although apparel image retrieval task has been well solved by current techniques to a certain degree, they were confined at the level of the numerical distance of the image feature vector. The computer still cannot understand what it has been seen at the fine-grained semantic level.

To enable the computer can really understand the fashion knowledge, researcher begins to combine the traditional method with deep learning. To mimic the human cognitive process, we introduce a framework, as shown in Fig. 4, as a reference for the roadmap for future.

In this framework, from the perspective of both fashion and AI, we divide the related tasks into three progressive levels. The first level is the fine-grained apparel recognition and semantic understanding based on the design feature of an apparel. To the public, people can easily distinguish the common categories (such as T-shirt, sweater, jacket, etc.) or design features (such as round neckline, V neckline, off-shoulder, etc.) based on their common knowledge. However, if they want to be a stylist, they should know how to choose the fashion items with some certain design features to form a good outfit and present a personal style as well. This task is not only requiring the apparel recognition in detail but also involving rules of mix and match which should be learnt by computer, which is the second level depicted in Fig. 6. The third level is the most advanced level in terms of creativity of computer on aesthetics. Specifically, on the basis of the foundation knowledge of the former two levels, computer can generate initial creative designs which can assist fashion designers for inspiring their new designs to certain degree.

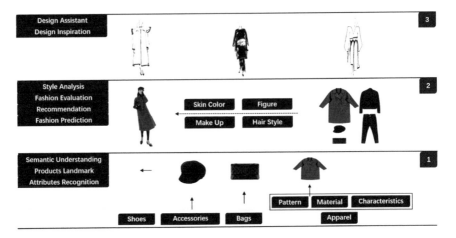

Fig. 6 Proposed framework of further research in fashion and AI

5 Conclusions

This paper reviews previous research studies of AI on the perspective of fashion. The fundamental technologies related to apparel parsing, attributes recognition, and fashion aesthetic understanding are summarized. Also, the available fashion dataset and current practical applications in the industry are also concluded. Moreover, we further propose a framework about the research direction of AI research on fashion as a reference for inspiring fashion and AI-related researchers.

Acknowledgements This paper was sponsored by the Hong Kong Polytechnic University.

References

1. Liu, S., Song, Z., Liu, G., Xu, C.: Street-to-shop: cross-scenario clothing retrieval via parts alignment and auxiliary set. In: Proc. CVPR (2012)
2. Reid, T.: Essays on the Intellectual Powers of Man. The M.I.T. Press, Cambridge, MA (1969)
3. James, S.: The Concept of the Aesthetic, Stanford Encyclopedia of Philosophy
4. Chen, H., Xu, Z.: Composite templates for cloth modeling and sketching. In: CVPR (2006)
5. Yang, M., Yu, K.: Real-time clothing recognition in surveillance videos. ICIP, 2937–2940 (2011)
6. Liu, S., Liang, X., Liu, L.: Fashion parsing with video context. IEEE Trans. Multimedia **17**(8), 1347–1358 (2015)
7. Wang, N., Ai, H.: Who blocks who: simultaneous clothing segmentation for grouping images. In: ICCV (2011)
8. Yamaguchi, K., Kiapour, M.H., Ortiz, L.E.: Parsing clothing in fashion photographs. In: CVPR (2012)
9. Yamaguchi, K., Kiapour, M.H., Berg, T.: Paper doll parsing: retrieving similar styles to parse clothing items. ICCV (2013)

10. Dong, J., Chen, Q., Xia, W.: A deformable mixture parsing model with parselets. In: ICCV, 3408–3415 (2013)
11. Yang, W., Luo, P., Lin, L.: Clothing co-parsing by joint image segmentation and labeling. In: CVPR, 3182–3189 (2014)
12. Liu, S., Feng, J., Domokos, C.: Fashion parsing with weak color-category labels. IEEE Trans. Multimedia 16(1), 253–265 (2014)
13. Simo-Serra, E., Fidler, S., Moreno-Noguer, F.: A high performance CRF model for clothes parsing. In: ACCV, 64–81 (2014)
14. Yamaguchi, K., Kiapour, M.: Retrieving similar styles to parse clothing. IEEE Trans. Pattern Anal. Mach. Intell. 37(5), 1028–1040 (2015)
15. Tangseng, P., Wu, Z., Yamaguchi, K.: Looking at outfit to parse clothing, arXiv preprint arxiv:1703.01386 (2017)
16. Neuberger, A., Alshan, E., Levi, G.: Learning fashion traits with label uncertainty. In: KDDW on ML meets fashion (2017)
17. Li, X., Yao, H., Sun, X.: Sparse representation based visual element analysis. In: ICIP, 657–660 (2011)
18. Wang, X., Zhang, T.: Clothes search in consumer photos via color matching and attribute learning. In: ACM, 1353–1356 (2011)
19. Fu, J., Wang, J., Li, Z.: Efficient clothing retrieval with semantic-preserving visual phrases. In: ACCV, 420–431 (2012)
20. Chen, Q., Li, J., Lu, G.: Clothing retrieval based on image bundled features. In: CCIS 2, 980–984 (2012)
21. Chen, H., Gallagher, A., Girod, B.: Describing clothing by semantic attributes. In: ECCV, 609–623 (2012)
22. Huang, C., Chen, S., Cheng, M.: A sparse-coding based approach to clothing image retrieval. In: ISPACS, 314–318 (2012)
23. Bossard, L., Dantone, M., Leistner, C.: Apparel classification with style. In: ACCV, 321–335 (2012)
24. Wei, D., Catherine W.: Style finder: fine-grained clothing style detection and retrieval. In: CVPRW, (2013)
25. Wang, X., Zhang, T., Tretter, D.: Personal clothing retrieval on photo collections by color and attributes. IEEE Trans. Multimedia 15(8), 2035–2045 (2013)
26. Kalantidis, Y., Kennedy, L., Li, J.: Getting the look: clothing recognition and segmentation for automatic product suggestions in everyday photos. In: ICMR, 105–112 (2013)
27. Kiapour M., Han, X., Lazebnik, S.: Where to buy it: matching street clothing photos in online shops. In: ICCV, 3343–3351 (2015)
28. Huang, J., Feris, R., Chen, Q.: Cross-domain image retrieval with a dual attribute-aware ranking network. In: ICCV (2015)
29. Mizuochi, M., Kanezaki, A., Harada, T.: Clothing retrieval based on local similarity with multiple images. In: ACM, 1165–1168 (2014)
30. Chen, Q., Huang, J., Feris, R.: Deep domain adaptation for describing people based on fine-grained clothing attributes. In: CVPR, 5315–5324 (2015)
31. Lin, K., Yang, H., Liu, K.: Rapid clothing retrieval via deep learning of binary codes and hierarchical search. In: ACM, 499–502 (2015)
32. Veit, A., Kovacs, B., Bell, S.: Learning visual clothing style with heterogeneous dyadic co-occurrences. In: ICCV, 4642–4650 (2015)
33. Yamaguchi, K., Okatani, T., Sudo, K.: Mix and match: joint model for clothing and attribute recognition. In: BMVC, 1(2) (2015)
34. Sun, G., Wu, X., Peng, Q.: Part-based clothing image annotation by visual neighbor retrieval. Neurocomputing 213, 115–124 (2016)
35. Simo-Serra, E., Ishikawa, H.: Fashion style in 128 floats: joint ranking and classification using weak data for feature extraction. In: CVPR, 298–307 (2016)
36. Vittayakorn, S., Umeda, T., Murasaki, K.: Automatic attribute discovery with neural activations. In: ECCV, 252–268 (2016)

37. Liu, Z., Yan, S., Luo, P.: Fashion landmark detection in the wild. In: ECCV, 229–245 (2016)
38. Liu, Z., Luo, P., Qiu, S.: Deepfashion: powering robust clothes recognition and retrieval with rich annotations. In: CVPR (2016)
39. Sengupta, B., Vasquez, E., Qian, Y.: Deep tensor encoding, arXiv preprint arxiv:1703.06324 (2017)
40. Chen, K., Luo, J.: When fashion meets big data: discriminative mining of best selling clothing features. In: WWWC, 15–22 (2017)
41. Laenen, K., Zoghbi, S., Moens, M.: Cross-modal search for fashion attributes. In: KDDW on ML meets fashion (2017)
42. Iwata, T., Wanatabe, S., Sawada, H.: Fashion coordinates recommender system using photographs from fashion magazines. In: IJCAI, 2262–2267 (2011)
43. Liu, S., Fen, J., Song, Z., Zhang, T., Lu, H.: Tell me what to wear! In: ACMMM (2012)
44. Liu, S., Liu, L., Yan, S.: Magic mirror: an intelligent fashion recommendation system. In ACPR 11–15 (2013)
45. Yamaguchi, K., Berg, T., Ortiz, L.: Chic or social: visual popularity analysis in online fashion networks. In: ACM, 773–776 (2014)
46. Jagadeesh, V., Piramuthu, R., Bhardwaj, A.: Large scale visual recommendations from street fashion images. In: ACM, 1925–1934 (2014)
47. Vittayakorn, S., Yamaguchi, K., Berg, A.: Runway to realway: visual analysis of fashion. In: WACV (2015)
48. Kiapour, M., Yamaguchi, K., Berg, A.: Hipster wars: discovering elements of fashion styles. In ECCV (2014)
49. Simo-Serra, E., Fidler, S., Moreno-Noguer, F., Urtasun, R.: Neuroaesthetics in fashion: modeling the perception of beauty. In: CVPR (2015)
50. Vaccaro, K., Shivakumar, S., Ding, Z.: The elements of fashion style. In: ASUIST (2016)
51. Al-Halah, Z., Stiefelhagen, R., Grauman, K.: Fashion forward: forecasting visual style in fashion. In: ICCV (2017)
52. Kang, W.C., Fang, C., Wang, Z.: Visually-aware fashion recommendation and design with generative image models, arXiv preprint arxiv:1711.02231 (2017)
53. Qian, Y., Giaccone, P., Sasdelli, M., et al.: Algorithmic clothing: hybrid recommendation, from street-style-to-shop, arXiv preprint arxiv:1705.09451 (2017)
54. Lin, Y., Wang, T.: Dress up like a stylist? learning from a user-generated fashion network[C]. In: KDDW on ML meets fashion (2017)
55. Han, X., Wu, Z., Jiang, Y.: Learning fashion compatibility with bidirectional lstms. In: ACM, 1078–1086 (2017)
56. He, R., McAuley, J.: Ups and downs: modeling the visual evolution of fashion trends with one-class collaborative filtering. In WWW, 507–517 (2016)
57. Garg, V., Banerjee, R.H., Anoop, K.R.: Sales potential: modelling shellability of visual aesthetics of fashion product. In: KDDW on ML meets fashion (2017)
58. Lee, D., Han, J., Chambourova, D.: Identifying fashion accounts in social networks. In: KDDW on ML meets fashion (2017)
59. Arora, S., Madvariya, A., Alok, D.: Deciphering fashion sensibility using community detection. In: KDDW on ML meets fashion (2017)
60. Mohammed Abdulla, G., Borar, S..: Size recommendation system for fashion e-commerce. In: KDDW on ML meets fashion (2017)
61. Lorbert, A., Ben-Zvi, N., Ciptadi, A.: Toward better reconstruction of style images with GANs. In: KDDW on ML meets fashion (2017)
62. Date, P.: Fashioning with networks: neural style transfer to design clothes. *CoRR* abs/1707.09899 (2017)
63. Zhu, S., Fidler, S., Urtasun, R.: Be your own prada: fashion synthesis with structural coherence, arXiv preprint arxiv:1710.07346 (2017)
64. Loni, B., Menendez, M., Georgescu, M.: Fashion-focused creative commons social dataset. In: ACM, 72–77 (2013)

65. Xiao, H., Rasul, K., Vollgraf, R.: Fashion-mnist: a novel image dataset for benchmarking machine learning algorithms, arXiv preprint arxiv:1708.07747 (2017)
66. Yu, Q., Liu, F., Song, Y.: Sketch me that shoe. In: CVPR, 799–807 (2016)
67. Yu, A., Grauman, K.: Fine-grained visual comparisons with local learning. In CVPR (2014)
68. Yu, A., Grauman, K.: Semantic jitter: dense supervision for visual comparisons via synthetic images. In: ICCV (2017)
69. Loni, B., Cheung, L.: Fashion 10000: an enriched social image dataset for fashion and clothing. In: ACM, 41–46 (2014)
70. Shao, M., Li, L., Fu, Y.: What do you do? occupation recognition in a photo via social context. In: ICCV, 3631–3638 (2013)
71. Gaur, A., Mikolajczyk, K.: ranking images based on aesthetic qualities. In: Proceedings of the International Conference on Pattern Recognition (ICPR) (2014)
72. Gaur, A., Mikolajczyk, K.: Aesthetics based assessment and ranking of fashion images. In: CVIU (2014)
73. Matzen, K., Bala, K., Snavely, N.: Streetstyle: exploring world-wide clothing styles from millions of photos, arXiv preprint arxiv:1706.01869 (2017)

Fashion Style Recognition with Graph-Based Deep Convolutional Neural Networks

Cheng Zhang, Xiaodong Yue, Wei Liu and Can Gao

Abstract Recognizing fashion styles of clothing from images plays an important role in the application scenarios of clothing retrieval and recommendation in E-commerce. Most existing works directly utilize the machine learning methods such as Deep Convolutional Neural Network (DCNN) to classify clothing images into different styles. However, these image classification methods are totally data-driven and neglect the domain issues of apparel fashion design. To tackle this problem, we propose a domain-driven clothing style recognition method in this paper, which involves both image classification and domain knowledge of fashion design. Specifically, we formulate the domain knowledge of design elements with the undirected graphs of clothing attributes and thereby build up a domain-driven fashion style classifier with Graph-Based DCNN. Synthesizing the classifications based on both clothing images and the graphs of design elements, we produce the final clothing style recognition results. The experiments on Deep Fashion database validate that the proposed clothing style recognition method can achieve more precise results than the traditional data-driven image classification methods.

Keywords Fashion style recognition · Deep convolutional neural networks

1 Introduction

Fashion style analysis of clothing images has become a fascinating domain in computer vision. Extensive efforts have been devoted to fashion style recognition because

X. Yue (✉)
Shanghai Institute for Advanced Communication and Data Science, Shanghai University, Shanghai, China
e-mail: yswantfly@shu.edu.cn

C. Zhang · X. Yue · W. Liu
School of Computer Engineering and Science, Shanghai University, Shanghai, China

C. Gao
Institute of Textiles and Clothing, The Hong Kong Polytechnic University, Hong Kong, China

© Springer Nature Switzerland AG 2019
W. K. Wong (ed.), *Artificial Intelligence on Fashion and Textiles*,
Advances in Intelligent Systems and Computing 849,
https://doi.org/10.1007/978-3-319-99695-0_32

269

of its potential values to textile/apparel industries and E-commerce such as clothing retrieval and recommendation [1–4]. But it is still challenging to recognize the fashion style from clothing images because of the ubiquitous variations of clothing items, which includes pose changing, scale variation, clothing deformation, and occlusions. Related works have been done to investigate the robust style recognition models to reduce the influence of clothing variations. One way is to detect the bounding boxes of the regions of clothing and human so that the clothes can be located [5]. Another way aims to tackle the problems of clothing deformations and occlusions and extracts the landmarks [6] which are the key points located at the functional regions of clothes to formulate more discriminative representations of clothes.

Most existing style recognition methods directly utilize machine learning methods, such as Deep Convolutional Neural Network (DCNN) to classify clothing images into different styles. However, in many cases, it is difficult to recognize clothing styles with only image data. For example, the clothing images having very similar texture features may belong to different fashion styles. Fashion style is actually a semantic description of clothing images and requires the interpretation of fashion design. The data-driven clothing image classification methods neglect the domain issues and may lead to inaccurate fashion style recognition results. Aiming to tackle this problem, we expect to involve the domain knowledge of fashion design in the process of clothing style recognition and thereby implement a domain-driven method for fashion style recognition.

Clothing attributes are semantic features to describe the fashion styles and design characterizations. For a fashion style, we formulate its domain knowledge of design elements with an undirected graph of the clothing attributes which co-occur in the clothes of this style. Based on the graphs of the clothing attributes of fashion styles, we adopt Graph-Based DCNN [7] to train one-versus-rest classifiers to recognize the fashion styles of clothing images. Moreover, we combine the domain-driven classification of Graph-Based DCNN and the data-driven classification of DCNN on clothing images to produce the final style recognition results. The contributions of this paperwork are summarized as follows. (1) Represent the domain knowledge of fashion design by graphs of clothing attributes. (2) Propose an ensemble approach for clothing style recognition domain-driven classification of Graph-Based DCNN and the data-driven classification of DCNN.

The rest of the paper is organized as follows. Section 2 presents the workflow of the proposed fashion style recognition method. In Sect. 3, we introduce the model of Graph-Based DCNN and its application in domain-driven style classification. In Sect. 4, experimental results validate the proposed method and the conclusions are given in Sect. 5.

2 Workflow

Figure 1 presents the workflow of the proposed fashion style recognition method, which are divided into two parts: data-driven recognition and domain-driven recog-

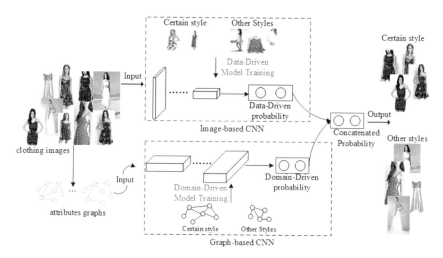

Fig. 1 Workflow of fashion style recognition

nition. The data-driven part directly trains DCNN on labeled clothing images and classifies the clothing images into fashion styles. The domain-driven part extracts the clothing attributes from the images of the same style to construct the undirected graphs of design elements and utilizes Graph-Based DCNN to build up another fashion style classifier. Concatenating the probabilities of both the classifications of DCNN and Graph-Based DCNN, we can determine the fashion style of a clothing image.

DCNN for data-driven recognition. The network structure is similar to VGG-16 [8] which has been demonstrated to be a powerful image classifier. Specifically, we maintain all the convolutional layers and pooling layers of VGG-16 and add one fully connected layer after the last pooling layer. The output layer contains two nodes to represent one certain style and the other ones, and uses softmax activation function to generate probabilities for style classification. Moreover, we pre-train the DCNN on ImageNet dataset and preserve the weights of convolutional layers and further train the fully connected layers using the bottleneck features output by the pre-trained network.

Graph-Based DCNN for domain-driven recognition. This part consists of three stages as illustrated in Fig. 2. In the first stage, the clothing attributes are extracted from the images of the same fashion style to construct the graphs of design elements. Second, the graphs are converted to feature vectors as the input of DCNN. In the final stage, a DCNN model is adopted to classify the feature vectors of graphs. The network consists of two convolutional layers and two fully connected layers. Both convolutional layers are one-dimensional. The conv1 has 16, 1×2 convolution kernels with a stride of 2 and ReLU activation. The conv2 has 8, 1×2 convolution kernels with a stride of 1 and ReLU activation. The first fully connected layer has 48

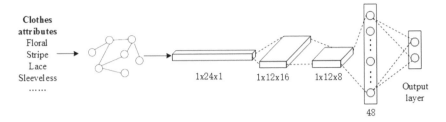

Fig. 2 Graph-based DCNN for domain-driven recognition

units. The output layer contains two nodes, which produce the probabilities belonging to one certain style and the others.

Concatenating Probabilities. To obtain the final recognition result, we need to concatenate the classification probabilities produced by two parts. Limiting the concatenated probability to [0, 1] and guaranteeing the sum of probabilities for all the styles is equal to 1, we adopt the following formula to concatenate probabilities.

$$\text{Prob}_{\text{result}} = \alpha \text{Prob}_{\text{image}} + (1 - \alpha)\text{Prob}_{\text{graph}}(0 < \alpha < 1) \tag{1}$$

where α is a weight, $\text{Prob}_{\text{image}}$ and $\text{Prob}_{\text{graph}}$ denotes the output probabilities of DCNN and Graph-Based DCNN, respectively, and $\text{Prob}_{\text{result}}$ is the final probability for style recognition.

3   Clothing Style Recognition with Graph-Based DCNN

Analogous to DCNN that operate on locally connected regions of images, the Graph-Based DCNN constructs locally connected neighborhoods from the input graphs. These neighborhoods are generated efficiently and serve as the receptive fields of a convolutional architecture, which facilitate DCNN to learn effective graph representation [8]. The key of Graph-Based DCNN is to determine the sequences of graph nodes to create neighborhoods and construct a unique mapping from the graph to a feature vector. The nodes should be properly positioned in the vector to maintain their structural roles in the graph neighborhoods.

To implement a domain-driven classifier with Graph-Based DCNN, we construct the undirected graphs of clothing attributes to represent the domain knowledge of fashion style. In the graph, each node represents one attribute of clothes. An edge connecting with two nodes denotes that the two attributes have strong co-occurrence relationship. To evaluate the strength of the relationship between two attributes, we construct a co-occurrence matrix of clothing attributes. An edge is added to connect with two nodes if the value in the matrix is larger than the threshold (set 10 as default). Figure 3 illustrates the co-occurrence matrix and the undirected graph of a rose style

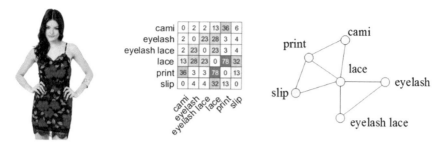

Fig. 3 Graph of a rose style dress

dress, which has 6 attributes. We can find that the fashion style of rose consists of the design elements of cami, lace, eyelash, slip and etc.

Referring to the inputs of Graph-Based DCNN, it is required to convert the constructed graphs to vectors. First, a fixed-length node sequence is determined according to the node centrality which is measured by node degree. Next, for each identified node, a receptive field is constructed through a breadth-first search. Finally, the receptive fields are converted into vector space representation according to the nodes centrality and co-occurrence matrix.

4   Experimental Results

DeepFashion is a large-scale clothes dataset to the community [1]. Each image in this dataset is labeled with 1000 descriptive attributes. We select 4040 dress images with any one of the four style labels from DeepFashion. From the 4040 images, we select a subset of 3002 images which with at least two attributes to construct graphs. For each of the style, we train a DCNN on clothing images and also build up a Graph-Based DCNN on clothing attribute graphs. The test dataset for each style consists of 180 images. To validate the proposed style recognition method, we first compare the performance of data-driven DCNN with the domain-driven Graph-Based DCNN and further overall evaluate the proposed method through comparing with other classic style recognition methods. We employee the measures of accuracy, precision, and recall to evaluate the performances of style recognition methods.

In Fig. 4, we can find that the proposed domain-driven style recognition method achieves higher accuracy, precision, and recall rate on average comparing with the data-driven method (+5.9, +4.3, and +6.8). Figure 5 shows the comparison of the recognition results obtained by our method and RestNet50, VGG-16, and InceptionResNetV2 [9]. We can also find that our method achieves the best performance among all the methods. The experimental results validate that involving the graphs of the design elements as domain knowledge is helpful to improve the precision of clothing style recognition.

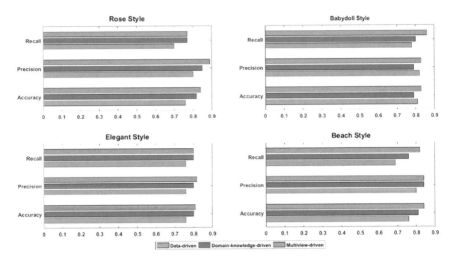

Fig. 4 Comparison of data-driven and domain-driven style recognition

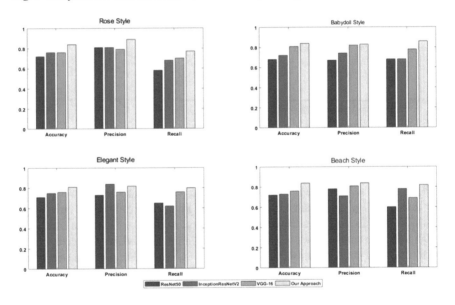

Fig. 5 Overall evaluation of style recognition methods

5 Conclusions

In this work, we propose a domain-driven clothing style recognition method with Graph-Based DCNN. The proposed method constructs the undirected graphs of attributes of clothes as domain knowledge of fashion design and involves the domain knowledge into the style recognition process. The experiments on DeepFashion dataset validate the proposed domain-driven style recognition method.

Acknowledgements This work reported here was financially supported by the National Natural Science Foundation of China (Grant No. 61573235).

References

1. Liu, Z., Luo P., Qiu, S., Wang, X., Tang, X.: Deepfashion: powering robust clothes recognition and retrieval with rich annotations. In: Proceedings of the IEEE Conference on Computer Vision and Pattern Recognition, pp. 1096–1104 (2016)
2. Huang, J., Feris, R., Chen, Q., Yan, S.: Cross-domain image retrieval with a dual attribute-aware ranking network. In: IEEE International Conference on Computer Vision, pp. 1062–1070 (2015)
3. Xiaodan, L., Liang, L., Wei, Y., Ping, L., Junshi, H., Yan, S.: Clothes co-parsing via joint image segmentation and labeling with application to clothing retrieval. IEEE Trans. Multimedia **18**(6), 1175–1186 (2016)
4. Qian, Y., Giaccone, P., Sasdelli, M., Vasquez, E., Sengupta, B.: Algorithmic clothing: hybrid recommendation, from street-style-to-shop (2017)
5. Hadi Kiapour, M., Han, X., Lazebnik, S., Berg, A.C., Berg, T.L.: Where to buy it: matching street clothing photos in online shops. In: IEEE International Conference on Computer Vision, pp. 3343–3351 (2015)
6. Liu, Z., Yan, S., Luo, P., Wang, X., Tang, X.: Fashion landmark detection in the wild. In: European Conference on Computer Vision, pp. 229–245 (2016)
7. Niepert, M., Ahmed, M., Kutzkov, K.: Learning convolutional neural networks for graphs. In: International conference on machine learning, pp. 2014–2023 (2016)
8. Simonyan, K., Zisserman, A.: Very deep convolutional networks for large-scale image recognition. Computer Science (2014)
9. Szegedy, C., Ioffe, S., Vanhoucke, V., Alemi, A.: Inception-v4, inception-resnet and the impact of residual connections on learning. AAAI (2017)

Fabric Defect Detection with Cartoon–Texture Decomposition

Ying Lv, Xiaodong Yue, Qiang Chen and Meiqian Wang

Abstract Automatic fabric defect detection plays an important role in textile industry. Most existing works utilize machine leaning methods to classify the fabric images with defects, however, because fabric defects are generally diverse and obscure. It is difficult to precisely identify the defects by direct image classifications. Aiming to tackle this problem, in this paper, we propose a two-stage method for automatic fabric defect detection. First, we utilize cartoon–texture decomposition to extract the features of textile structures from fabric images. Second, based on the features of cartoon textures, we build up a classifier with Deep Convolutional Neural Networks (DCNN) to distinguish the image regions containing defects, i.e., the regions of abnormal feature representation. Experimental results validate that the proposed method can precisely recognize the fabric defects and achieve good performances on the fabric images of various kinds of textiles.

Keywords Fabric defect detection · Cartoon–texture decomposition
Deep convolutional neural networks

1 Introduction

In the textile field, the weaving process can produce fabric defects which include width inconsistencies, hairiness, slubs, broken ends, etc. [1]. Automatic fabric defect detection is helpful to improve the textile quality and reduce the costs of human and material resources. The existing methods of automatic fabric defect detection are summarized as the following kinds: textile structure methods, statistical meth-

X. Yue (✉)
Shanghai Institute for Advanced Communication and Data Science,
Shanghai University, Shanghai, China
e-mail: yswantfly@shu.edu.cn

Y. Lv · X. Yue · Q. Chen · M. Wang
School of Computer Engineering and Science,
Shanghai University, Shanghai, China

© Springer Nature Switzerland AG 2019
W. K. Wong (ed.), *Artificial Intelligence on Fashion and Textiles*,
Advances in Intelligent Systems and Computing 849,
https://doi.org/10.1007/978-3-319-99695-0_33

ods, spectral methods, model-based detection, and machine learning approaches [2]. However, because fabric defects are generally diverse and obscure, it is still challenging to efficiently and precisely distinguish defects from fabric images [3]. Especially for the fabric with complex texture structures, traditional detection methods may miss the inconspicuous defects and lead to inaccurate results [4].

In this paper, we expect to improve the fabric defect detection based on textile structure analysis. Textile structures can be modeled as a composition of texture primitives [3, 5]. The analysis of structural textures mainly consists of two steps: texture extraction and placement inference [6]. The detection based on texture analysis facilitates recognizing defects in the fabrics of regular textile structures. But there are following limitations of texture-based defect detection methods: (1) it is difficult to extract the structural textures of complex weave patterns and (2) the identification and localization of defective regions are not precise.

Aiming to handle the problems above, we propose a novel texture-based method for fabric defect detection, which integrates cartoon–texture decomposition [7] and Deep Convolutional Neural Network (DCNN). Specifically, the proposed method consists of two stages. First, we utilize cartoon–texture decomposition to extract the texture features of textile structures. Second, based on the features of cartoon–textures, we build up a classifier with Deep Convolutional Neural Networks (DCNN) to distinguish the image regions containing defects. Cartoon textures are effective to represent diverse textile structures and in the meantime sensitive to the fabric defects, which is helpful to produce robust detection results.

The rest of this paper is organized as follows. Section 2 introduces the workflow of the proposed method which includes structural texture extraction with cartoon—texture decomposition and defective region detection based on DCNN. Section 3 presents the experimental results and validates the effectiveness of the proposed method for fabric defect detection. The work conclusion is given in Sect. 4.

2 Defect Detection with Cartoon–Texture Decomposition

2.1 Workflow of Defect Detection

The defective area is the heterogeneous texture. It has significant differences from regular texture of normal fabric. Thus, the defective area can be identified more accurately on fabric texture part. Based on the above facts, a complete overview of the fabric defect inspection method is presented in Fig. 1.

As shown in Fig. 1, the workflow of fabric defect detection includes two stages.

(1) **Texture Extraction**. We can obtain the fabric texture part and cartoon part by cartoon–texture decomposition algorithm. The cartoon part is a simplified approximation of the image, which includes the main feature of the image. The texture part is oscillatory component or sampling statistics, which includes the texture of image and noise.

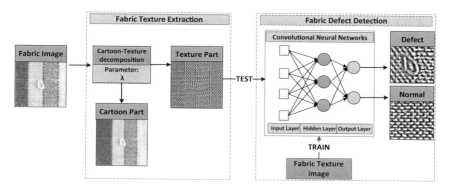

Fig. 1 The workflow of fabric defect detection

(2) **Defect Prediction**. The classifier used in our method is DCNN, which is known as a discriminant model. Texture parts of normal and defect images are used to train the network. When we apply the trained DCNN to distinguish the defective areas, it outputs the probabilities of an input fabric texture belonging to defective and normal part. Therefore, we define that the input texture belongs to the category with higher probability.

2.2 Fabric Texture Extraction with Cartoon–Texture Decomposition

In the cartoon–texture decomposition algorithm, a fabric image f is defined on a continuous domain in a finite set of pixels. The cartoon–texture decomposition algorithm makes the fabric image f into the sum of cartoon part u and texture part v based on Meyer's models. We focus on the two components: the cartoon u is piecewise-smooth component of f that has the global structural information, and the texture v represents the locally patterned oscillatory component of f. Due to obvious warp yarn and weft yarn in weave image, the oscillatory nature of fabric is noticeable. And in contrast with simple low-pass filters, cartoon–texture decomposition methods permit sharp edges of complex geometry in the cartoon part. So, the fabric texture and the defective area will be contained in texture part v and identification defective areas in texture part are easier than in the original image.

As mentioned above, we estimate the cartoon part u and the texture part v of fabric image f by cartoon–texture decomposition methods. The image f can be decomposed as

$$f = u + v \tag{1}$$

A general variational framework of the cartoon–texture decomposition is solving energy minimization problem.

$$\inf_{(u,v)\in X_1\times X_2}\{\lambda F_1(u)+F_2(v):f=u+v\} \tag{2}$$

where $F_1, F_2 \geq 0$ are functionals and X_1, X_2 are function spaces or distributions such that any function pair $(u, v) \in X_1 \times X_2$ fulfills $F_1(u) < +\infty$ and $F_2(v) < +\infty$, $\lambda > 0$ is a trade-off between $F_1(u)$ and $F_2(v)$. By selecting appropriate functionals for F_1 favors piecewise-smooth functions u satisfies $F_1(u) \ll F_1(v)$ and F_2 favors oscillatory functions v satisfie, $F_2(u) \ll F_2(v)$. The fabric image f can be decomposed into cartoon part u and texture part v by solving the above optimization problem [8].

There have been a lot of researches in designing functionals F_1 and F_2. In this paper, we will apply the nonlinear filtering cartoon–texture decomposition method of [7]. It provides faster and better decomposition strategies for a cartoon and texture component.

2.3 Classification with DCNN

In this section, we adopt the DCNN for the fabric defect detection in weave texture image. The main idea is to learn a hierarchy of feature detectors and train a nonlinear classifier to identify defective areas in the fabric texture part.

Figure 2 shows the framework of our networks. We partition a fabric texture image into 400 patches. Such patches are as the input of neural network and the patches of normal and the defect train the networks. The convolutions layers can get the abstract texture features by convolution kernels. The abstract texture features are more expressive in fabric image. We apply the Rectified Linear Units (**ReLUs**) as activation in the convolutional and fully connected layers. The loss function is negative log-likelihood. And we optimize the parameters of networks by Stochastic Gradient Descent (SGD). We use the logistic regression with softmax to calculate the probability on defect and normal, as defined.

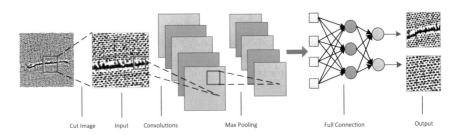

Cut Image   Input   Convolutions   Max Pooling   Full Connection   Output

Fig. 2 The process of cartoon–texture decompose algorithm

$$P(y = i|x, W_1, \ldots, W_c, b_1, \ldots, b_c) = \frac{e^{w_i x + b_i}}{\sum_{j=1}^{C} e^{w_j x + b_j}} \tag{3}$$

where x is the output of the fully connected second layer, W_i and b_i are the weights parameter ith layer and offset of ith layer, and C is the number of classes. The outputs class with the max probability is the predicted class, which is expresses as follows ($\hat{y}$ is predicted class):

$$\hat{y} = \arg\max_i P(y = i|x, W_1, \ldots, W_c, b_1, \ldots, b_c) \tag{4}$$

As introduced above, we can detect fabric defective areas by the DCNN.

3   Experimental Results

We implement two experiments to validate the proposed method for the fabric defect detection. The datasets have six types of fabric defect (including weft bow, oil stain, blot, cockled yarn, loom fly, and dislocation structure) that are shown in Fig. 3.

First, to validate, the cartoon–texture decomposition algorithm can obtain accurately the texture part of fabric and the defect is more significant in texture part. We extract the texture part by cartoon–texture decomposition in Fig. 4. To compare (a) with (b), we can accurately extract the fabric structure (including yarn, weft, defective areas, and noise). And to compare (c) with (d), we can observe that the defective areas are more distinct. Thus, the cartoon–texture decomposition is suited to extract the fabric texture.

The second experiment overall evaluates the performance of defect detection with cartoon–texture decomposition in fabric texture part through comparing with defect detection in original fabric image. We partition a fabric texture image into 400 patches. And we use the five types of fabric defect images to train the DCNN and one type of fabric defect images for testing. The results are shown in Fig. 5.

In the same experimental environment, we use the DCNN to discern fabric defect in the original image. The result is shown in Fig. 6. To compare Fig. 5 with Fig. 6, the result indicates that the performance of method is more accuracy in the texture image.

| (a) | (b) | (c) | (d) | (e) | (f) |

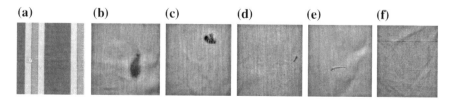

Fig. 3   **a** Weft bow, **b** oil stain, **c** blot, **d** cockled yarn, **e** loom fly, **f** double yarn

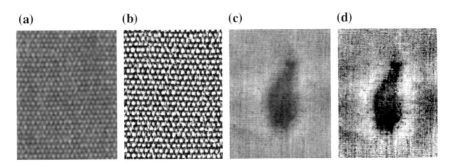

Fig. 4 **a, c** Original image, **b, d** texture of original image

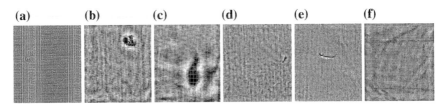

Fig. 5 Fabric defect detection on texture part

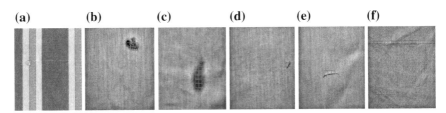

Fig. 6 Fabric defect detection on original image

4 Conclusion

Most existing structure approaches of fabric defect detection have the drawbacks in the feature extraction of texture and accuracy of fabric defect recognition. To solve these problems, in this paper, we propose a texture extraction method based on cartoon–texture decomposition. And we use the DCNN to detect fabric defect in the texture part. Experimental results validate that the fabric texture extraction based the cartoon–texture decomposed method is effective for fabric defect detection. Our future work will focus on identifying the types of defective areas.

Acknowledgements This work reported here was financially supported by the National Natural Science Foundation of China (Grant No. 61573235).

References

1. Mahajan, P.M., Kolhe, S.R., Patil, P.M.: A review of automatic fabric defect detection techniques. Adv. Comput. Res. **1**(2), 18–29 (2009)
2. Hanbay, K., Talu, M.F., Ömer, F.: Fabric defect detection systems and methods—A systematic literature review. Optik—Int. J. Light and Electron Optics **127**(24), 11960–11973 (2016)
3. Ngan, H.Y.T., Pang, G.K.H., Yung, N.H.C.: Automated fabric defect detection—A review. Image Vis. Comput. **29**(7), 442–458 (2011)
4. Kumar, A.: Computer-vision-based fabric defect detection: a survey. IEEE Trans. Industr. Electron. **55**(1), 348–363 (2008)
5. Li, Y., Zhao, W., Pan, J.: Deformable patterned fabric defect detection with fisher criterion-based deep learning. IEEE Trans. Autom. Sci. Eng. **14**(2), 1256–1264 (2017)
6. Sayed, M.S.: Robust fabric defect detection algorithm using entropy filtering and minimum error thresholding. In: IEEE ISCAS. IEEE, pp. 1–4 (2017)
7. Buades, A., Le, T.M., Morel, J.M., et al.: Fast cartoon+texture image filters. IEEE Trans. Image Process. **19**(8), 1978–1986 (2010)
8. Meyer, Y.: Oscillating Patterns in Image Processing and Nonlinear Evolution Equations: The Fifteenth Dean Jacqueline B. Lewis Memorial Lectures. University Lecture Series (2001)

Fabric Texture Removal with Deep Convolutional Neural Networks

Li Hou, Xiaodong Yue, Xiao Xiao and Wei Xu

Abstract In this paper, we propose a neural network based on Deep Convolutional Neural Network (DCNN) to remove fabric textures from scanned images. Different from the traditional DCNN performed on original images, the proposed network focuses on extracting texture structures and utilizes the texture layers of fabric images for model training. To achieve the precise extraction of fabric textures, the proposed model adopts a network architecture which is inspired by the deep residual network (ResNet). Experiment results on multiple kinds of fabric images validate that the proposed network is effective to remove fabric textures and achieves better performances than other kinds of denoising methods.

Keywords Fabric texture removal · Deep convolutional neural network

1 Introduction

In fabric industrial applications, high-quality scanning images of fabrics are very important for fashion design and textile printing and dyeing. This requires to remove the fabric textures formed in textile process from the scanned images. Because of the complexity of the fabric textures, it is difficult for scanning devices to directly distinguish and remove the texture patterns. Considering fabric textures as noise, the image denoising methods, such as frequency filters and sparse coding, have been used to remove the textures from the scanned fabric images [1–4]. However, the frequency filters aim to remove the noise of high frequencies rather than structural textures and may cause the information loss of image contents. Sparse coding methods specialize

X. Yue (✉)
Shanghai Institute for Advanced Communication and Data Science,
Shanghai University, Shanghai, China
e-mail: yswantfly@shu.edu.cn

L. Hou · X. Yue · X. Xiao · W. Xu
School of Computer Engineering and Science,
Shanghai University, Shanghai, China

© Springer Nature Switzerland AG 2019
W. K. Wong (ed.), *Artificial Intelligence on Fashion and Textiles*,
Advances in Intelligent Systems and Computing 849,
https://doi.org/10.1007/978-3-319-99695-0_34

in denoising grayscale images and often lead to unclear texture removal on colorful fabric images.

Aiming to tackle the problem above, we expect to design a model to remove the fabric textures from scanned images, and in the meantime preserve the content details, which facilitate clothing design and printing. To achieve this, we implement a neural network based on Deep Convolutional Neural Network (DCNN) to extract and remove fabric textures. The proposed network focuses on modeling the texture structures and thereby facilitates the precise extraction of fabric textures. The basic idea of the proposed network is to reconstruct the texture layer of a scanned fabric image through regression and adopt DCNN to learn the regression function. Different from the traditional DCNN performed on original images, we adopt only the texture layers of images to train the networks for regression. To achieve this, we implement the network architecture similar to the deep residual network (ResNet) [5, 6] and use the negative residual as the output of the network. Comparing with the original images, the image residuals have significant range reduction and are helpful to improve the learning of regression function with DCNN [7].

The rest of the paper is organized as follows. Section 2 introduce the proposed network for fabric texture removal/extraction. In Sect. 3, experimental results validate the proposed model and the conclusions are given in Sect. 4.

2   Fabric Texture Removal with DCNN

2.1   Workflow

The framework of the model is shown in Fig. 1. As discussed below, we define the texture layer and the negative residual to be the input and output of parameter layers. In the process of experiment, we find that it will get better results if the difference between the input images and ground truth is only texture structure. Thus we use the manual method to make the ground truth instead of using original paintings. After training, the network can be used to remove texture from any other scanned images of cloth.

2.2   Texture Layer and Residual

We denote the input image with texture and the ground truth as X and Y. A goal of training a DCNN architecture $h(X)$ is to minimize the objective function

$$L = \sum_{i} \|h(X_i) - Y_i\|_F^2, \tag{1}$$

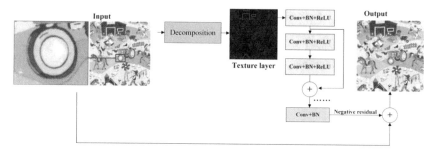

Fig. 1 Framework of fabric texture removal with DCNN

Fig. 2 Result of direct
DCNN and our network

where F is Frobenius norm. However, we find that the results by the direct DCNN has some color shift, as we can see in Fig. 2. In general, we observed that the learned regression function underfits when directly trained on the image domain. It is because the mapping range covers all possible pixel values, which makes it hard to learn the regression function well [5]. Meanwhile, training a deep network directly on images suffers from gradient vanishing even when using regularization methods.

Since the complex mapping leads to the poor results, we must improve the learning process by reducing the solution space. When compared to the ground truth Y, we observe that the residual of Y-X has a significant range reduction in pixel values. So that our network try to incorporates this idea. We also find that most values of Y-X tend to be negative, so we refer to this as "negative residual mapping". A modified objective can be written as

$$L = \sum_i \|h(X_i) + X_i - Y_i\|_F^2. \tag{2}$$

Figure 2 also shows the result on ResNet architectures that is improved by using neg-mapping. We use SSIM for a quantitative evaluation.

Since reducing the solution space helps improve the learning process, we try to use the sparsity of images to increase the performance even further. In contrast to the original ResNet, we use the texture layer as input to the parameter layers. We can make the scanned image as

$$X = X_{\text{texture}} + X_{\text{base}}, \tag{3}$$

where the subscript 'texture' denotes the texture layer, and 'base' denotes the base layer. In our work we use the low-pass filtering of X to obtain base layer. After that, the texture layer is $X_{\text{texture}} = X - X_{\text{base}}$. The texture layer is sparser than the scanned image. Therefore, the effective mapping in texture layer is from smaller subsets of Neg-mapping. This indicates that the solution space has shrunk and so network performance should be improved [6].

2.3   Objective Function and Network Architecture

Based on the previous discussion, we define the objective function to be,

$$L = \sum_{i=1}^{N} \left\| f(X_{i,\text{texture}}, W, b) + X_i - Y_i \right\|_F^2, \tag{4}$$

where N is the number of training images, $f(\cdot)$ is ResNet, W and b are network parameters that need to be learned. For X_{texture}, we first use a low-pass filter to split X into base and texture layers.

Our network structure can be expressed as,

$$
\begin{aligned}
X_{\text{texture}}^0 &= X - X_{\text{base}}, \\
X_{\text{texture}}^1 &= \sigma(BN(W^1 * X_{\text{texture}}^0 + b^1)), \\
X_{\text{texture}}^{2l} &= \sigma(BN(W^{2l} * X_{\text{texture}}^{2l-1} + b^{2l})), \\
X_{\text{texture}}^{2l+1} &= \sigma(BN(W^{2l+1} * X_{\text{texture}}^{2l} + b^{2l+1})) + X_{\text{texture}}^{2l-1}, \\
Y_{\text{approx}} &= BN(W^L * X_{\text{texture}}^{L-1} + b^L) + X,
\end{aligned}
\tag{5}
$$

where $l = 1, \ldots, \frac{L-2}{2}$ with L the total number of layers, $*$ indicates the convolution operation, W contains weights and b biases, $BN(\cdot)$ indicates batch normalization to alleviate internal covariate shift, $\sigma(\cdot)$ is a Rectified Linear Unit (ReLU) for nonlinearity. In this network, all pooling operations are removed to preserve spatial information.

3   Experimental Results

We use the result of our network to compare with the manual result, using SSIM as evaluating indicator. For comparison, we also show the SSIM of results based on the method of frequency domain. Table 1 shows the layer and parameters of the network.

Figure 3 shows the result on the fabric with complex texture. As we can see from the zoomed region, although the origin pattern is intricate, our network output still can remove the texture with keeping the pattern clear. Figure 4 shows the result of other method which is mentioned before. As we can see from Fig. 4, method using

Table 1 The layer and parameters of DCNN for fabric texture removal

Epoch: 100	Initial learning rate: 0.1		Depth: 26
Layer	1st	2nd–25th	26th
Structure	Conv+BN+ReLU	Conv+BN+ReLU	Conv+BN
Kernel size	3×3	3×3	3×3
Input/output channel	3/16	16/16	16/3

Fig. 3 Our results on a test image 1

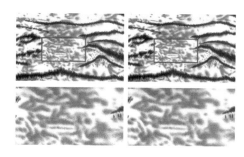

Fig. 4 Result on method of frequency domain and direct DCNN

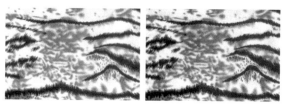

Fig. 5 Two results on test image 2 and 3

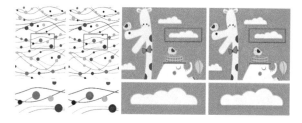

low-level features get bad result because of the difficulty to distinguish the texture and detail of image. This leads to the texture remain. Result on direct DCNN still suffers the color shift. Moreover, the damage to image structure is severe.

Figure 5 shows the results on scanned image of fabric which have different style with image in Fig. 3. As shown, the result on image with regular and clean pattern is the best. On the image with cartoon style, which is bright in color, our output can still remove the texture and preserve the origin color.

We give the quantitative evaluation using SSIM on 100 images in Table 2. The amount of test data is small because our ground truth is generated artificially which is very time consuming. A Higher SSIM indicate better result. As we can see in Table 2, the output from method of frequency domain result in a lower SSIM. In some test

Table 2 Quantitative measurement results using SSIM on test images

Images	Ground truth	Texture image	Frequency domain	Ours
Test image 1	1	0.61	0.65	0.86
Test image 2	1	0.54	0.62	0.95
Test image 3	1	0.74	0.70	0.83
100 test images	1	0.60	0.57	0.81

images, the SSIM index for method of frequency domain is even lower than the texture images, which means the low-level feature method removes the non-texture part. On scanned images with different style, our method result in a better SSIM.

4 Conclusion

In this work, we propose an image denoising method applied on scanned image of fabric based on DCNN network. Different from traditional DCNN method, the proposed method using the texture layer as input and the residual as the output of network. The sparseness of the input and output optimize the training process of network. The experiments on our data show the better result compared with other methods.

Acknowledgements This work reported here was financially supported by the National Natural Science Foundation of China (Grant No. 61573235).

References

1. Cornelis, B., Yang, H., Goodfriend, A., Ocon, N., Lu, J., Daubechies, I.: Removal of canvas patterns in digital acquisitions of paintings. IEEE Trans. Image Process. **26**(1), 160–171 (2016)
2. S, Li., Huang, D.S., Zheng, C.H., Sun, Z.L. Letters: noise removal using a novel non-negative sparse coding shrinkage technique. Neurocomputing **69**(7), 874–877 (2006)
3. Chen, Y., Cao, X., Zhao, Q., Meng, D., Xu, Z., et al. Denoising hyperspectral image with non-i.i.d. noise structure. IEEE Trans. Cybern. **48**(3), 1054–1066 (2017)
4. Zhang, K., Zuo, W., Chen, Y., Meng, D., Zhang, L.: Beyond a gaussian denoiser: residual learning of deep CNN for image denoising. IEEE Trans. Image Process. **26**(7), 3142–3155 (2017)
5. Fu, X., Huang, J., Zeng, D., Huang, Y., Ding, X., Paisley, J. Removing rain from single images via a deep detail network. In: CVPR, pp. 1715–1723 (2017)
6. He, K., Zhang, X., Ren, S., Sun, J. Deep residual learning for image recognition. In: CVPR, pp. 770–778 (2015)
7. Simonyan, K., Zisserman, A. Very deep convolutional networks for large-scale image recognition. In: Computer Science (2014)

Optimal Gabor Filtering for the Inspection of Striped Fabric

Le Tong, Xiaoping Zhou, Jiajun Wen and Can Gao

Abstract As an important part of products' quality control, automatic fabric inspection has attracted much attention in the past. Compared with manual inspection, automatic inspection can achieve not only more accurate detection results but also a higher efficiency. With the diversified fabric texture and patterns, it is very necessary to develop distinctive detection methods for different types of fabric. In this paper, based on optimal Gabor filters, a novel defect detection model is proposed to address the inspection of striped fabric, which is commonly used in our daily dresses. In the framework of the detection model, Gabor filters perpendicular to the stripe pattern are optimized to minimize the variance of the image but enhance the features of defects. Thereafter, an adaptive thresholding is set to accurately segment the defective image area. The evaluation of the proposed detection model is conducted using samples of the TILDA database. It is revealed that the common fabric defects as well as the pattern variants could be successfully detected through the proposed detection model.

Keywords Fabric inspection · Optimal Gabor transformation · Stripe fabric

L. Tong · X. Zhou
College of Information, Mechanical and Electrical Engineering,
Shanghai Normal University, Shanghai 200072, China
e-mail: tongle@shnu.edu.cn

J. Wen (✉) · C. Gao
College of Computer Science and Software Engineering,
Shenzhen University, Shenzhen 518055, China
e-mail: enjoy_world@163.com

J. Wen · C. Gao
Institute of Textile and Clothing, The Hong Kong Polytechnic University,
Kowloon, Hong Kong

J. Wen
The Hong Kong Polytechnic University Shenzhen Research Institute, Shenzhen 518055, China

© Springer Nature Switzerland AG 2019
W. K. Wong (ed.), *Artificial Intelligence on Fashion and Textiles*,
Advances in Intelligent Systems and Computing 849,
https://doi.org/10.1007/978-3-319-99695-0_35

1   Introduction

In the textile and apparel enterprises, fabric quality inspection plays an important role in controlling cost and boosting profit, because locating possible fabric defects before cutting is beneficial for reducing raw material waste and ensuring textile product quality. According to the related survey [1], the accuracy of the traditional manual inspection, which relies on visual checking by experienced operators, is only approximately 60–75%. Taking advantages of digital imaging devices and computer vision techniques, automatic fabric defect detection system can simulate human visual system, and provide intelligent solutions for defect detection without human intervention, which is beneficial for improving detection accuracy and efficiency. Therefore, the investigation of computer vision-based fabric defect detection methods is of great significance.

Fabric defects may exhibit a variety of forms because of the differences in the weaving structures. Defects on the surface of uniform texture fabric (plain and twill fabric in solid color) usually exhibit as prominent edges. The case on patterned fabrics (e.g., print fabric) is much more complicated. In addition to the common defects that are likely to be confused with the patterns, defects may occur on the pattern primitive unit itself or its arrangement. Moreover, there are various kinds of fabric patterns, each of which has different characteristics. Therefore, to effectively address the automatic fabric defect detection problem in the textile and apparel industries, the most feasible solution is to develop different detection approaches for different fabric types.

Over the past two decades, a variety of feature extraction-based methods have been developed for the defect detection of uniform texture. The idea of these methods is to identify defects through extracting and comparing the features of normal or abnormal fabric texture. In [2], multiple statistical features of gray-level co-occurrence matrix were calculated to discriminate the defective and non-defective area. In [3], the log-likelihood map derived from the coefficients of hidden Markov tree models of images was used as the measurement of defective samples. Wen et al. [4] applied adaptive wavelet transformation to extract the intrinsic textural characteristics of non-defective fabric images. Tong et al. [5] proposed a fabric inspection model based on two horizontally and vertically optimized Gabor filters, which significantly improved the detection accuracy and efficiency. Although the above methods have achieved good performance in the inspection of plain and twill fabric, the detection results on patterned fabric are not satisfied. It is very hard to ensure the best discrimination between the defects and normal texture, when defects are likely to overlay the complex fabric patterns.

In recent years, some pattern analysis-based methods have emerged to address the inspection of patterned fabric. With the printed pattern regarded as the underlying lattice, some motif-based defect detection approaches employed the characteristics of the smallest repeated unit in the fabric image [6, 7]. Regarding defects as noise in the image, some sparse representation-based methods have been proposed. The defects are segmented through the comparison of the input image and its approximation [8, 9]. By applying probabilistic neural network and pulse coupled neural network

[10, 11], the correlations between local image patches and their neighborhood are designed as the inputs of network, while the outputs indicate whether an image patch is defective or not. Pattern analysis-based approaches usually cost amount of computational work, which makes the inspection efficiency of fabric with basic type of pattern such as striped fabric very low. Therefore, it is very necessary to develop distinctive detection methods according to the specific characteristics of the inspected fabric.

In this paper, we present a novel defect detection model based on optimal Gabor filter for the inspection of striped fabric, which represents the most basic type of patterned fabric. First, according to the angle features of sample images, a Gabor filter is optimized to maximally eliminate the response of stripe pattern in the feature space. Thereafter, an adaptive thresholding operation is set to outline the shape of defects.

2 Defect Detection Based on Optimal Gabor Filter

Figure 1 shows a few samples of non-defective and defective striped fabric. The variance of striped fabric is significantly larger than uniform textural fabric; thus, small defects do not cause perceivable changes to the statistical features of the entire image. Therefore, reducing the interference brought by stripe patterns in detecting defects is a key issue in the design of striped fabric inspection model.

Owing to the optimal joint localization in the spatial and frequency domains, Gabor transform has become a promising method for fabric defect detection. As shown in Eq. (1), the freedom in the selection of parameters of a Gabor filter permits to select the response of object that we are interested in the filtered image. An imaginary Gabor filter which is designed to enhance the response of horizontal lines, will accordingly restrain the response of vertical lines. Therefore, an imaginary Gabor filter can be optimized to eliminate the normal stripe pattern in the first step, so that the features of pattern variants and common defects can be kept.

$$f(x, y) = \frac{1}{2\sigma_x \sigma_y} \exp\left[-\frac{1}{2}\left(\frac{x'^2}{\sigma_x'^2} + \frac{y'^2}{\sigma_y'^2}\right)\right] \cdot \exp(2\pi j f_0 x') \tag{1}$$

Fig. 1 Non-defective and defective samples of a striped fabric

2.1 Optimal Gabor Filtering Based on CoDE

The histogram of a striped fabric image indicates that there are more pixels with high gray-level intensities when the stripe pattern is more salient. Thus, in order to reduce the disturbance caused by the stripe pattern, the primary objective in the Gabor filter optimization is to minimize the mathematical expectation of the histogram of the edge image extracted from the filtered feature image by Sobel mask. For an image F with $M \times N$ pixels, the objective S can be formulated as

$$S = \sum_{i=0}^{255} i \frac{q_i}{M \times N} \tag{2}$$

where i denotes the value of gray levels, and q_i is the number of pixels under ith gray level in the edge image extracted from F. It makes the abrupt intensity changes caused by the stripe pattern significantly eliminated. Further, the variance of the filtered feature image denoted by σ should be as small as possible to ensure that the texture of the original fabric sample is as uniform as possible after Gabor filtering, which is beneficial for the robustness of defect detection. Therefore, the final objective of the Gabor optimization is

$$\text{Minimize} \quad J = \sigma \times \sum_{i=0}^{255} i \frac{q_i}{M \times N} \tag{3}$$

Given that the proposed optimization objective function contains many local optimal solutions, the evolutionary algorithm CoDE, which has shown superior performance in solving multi-modal problems is selected as the optimization algorithm to obtain the optimal set of parameters of Gabor filter. The orientation parameter is designed perpendicularly to the stripe pattern. Hence, three Gabor parameters should be optimized: the central frequency of Gabor filter f_0 together with the smoothing parameters σ_x and σ_y.

2.2 Adaptive Thresholding

According to the objective function Eq. (3), the energy response of the repeated stripe pattern will be minimum, and most of the image area is homogeneous, except from the defects. Therefore, s simple thresholding operation could distinguish defective pixels from homogeneous pixels in the feature image. In the filtered feature image, the area of normal fabric texture remains a small amount of filtering response, comparing with the defects. It is a great practice to adaptively achieve the local thresholding limit based on the pixel value distribution of the pixel's neighborhood block. In this paper, the binary image $B(x, y)$ which represents the features of defects could be obtained by

$$B(x, y) = \begin{cases} 1, & \text{if } F(x, y) > \varepsilon + k\sigma \\ 0, & \text{otherwise} \end{cases} \qquad (4)$$

where, $F(x, y)$ represents the filtered feature image; ε and σ denotes the gray-level mean value and variance within a local image block, respectively; k is a constant that practically determined.

3 Experiments

In this section, the performance of the proposed fabric defect detection model is validated on TILDA, in which 50 non-defective samples and 300 defective samples are included. In the database, pattern defects include compactor crease and pattern variants caused by mechanical fault; and common defects include chafed yarn, flying yarn and color mark.

Figures 2 and 3 demonstrate several representative detection results using the proposed defect detection model. A non-defective sample is selected as the reference to optimize the Gabor filter at the beginning of the inspection of each type of fabric. Images in the second column show the optimal filtering response, while the binary images in the third column show the final detection results. The overall experimental results are summarized in Table 1.

Figure 2 demonstrates the detection results of pattern variants. The first sample is the variation of repeated stripe pattern which is usually caused by the restart of

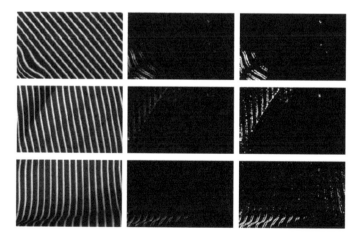

Fig. 2 Detection results of pattern variants

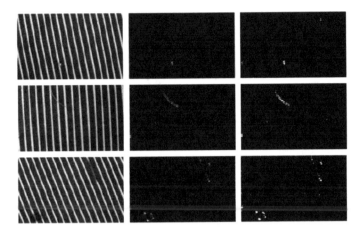

Fig. 3 Detection results of common fabric defects

Table 1 Performance of the proposed detection model on striped fabric samples of TILDA

	Sensitivity (%)	Detection accuracy (%)
The proposed model	90.0	87.1

the weaving machine or a malfunction in the weaving tensor. The last two samples are fabric crease. All these defects are successfully detected. The edges of the incorrectly arranged stripe are not eliminated in the filtering response in the first step of inspection, because these misarranged stripes deviate from the standard orientation of repeated patterns.

The detection results of three common defects on the striped fabric are shown in Fig. 3. The first two are structural defects related to weaving texture. Although there are some overlaps between these two defects and the stripe pattern, they are accurately outlined. The last one is a tonal defect that only alters the local intensity value of the fabric but not the weaving structure. The overlapped part of the defect is successfully detected, which further verifies that the proposed detection model is sensitive to most types of fabric defects.

4 Conclusion

This paper has tackled the inspection problem of striped fabric by proposing an optimal Gabor filter-based defect detection model. The Gabor filter was optimized to minimize the filtered response of repeated stripe pattern, which is beneficial for increasing the discrimination between defects and fabric background. The performance of the detection model is evaluated by extensive experiments on TILDA database. The experimental results on misarranged and crease defects indicate that the proposed detection model is effective on any directional pattern defects because of the selective orientation of Gabor filters. Accurate detection results on common fabric defects also prove that the proposed model performs well in detecting most types of fabric defects while reducing the interference of stripe pattern.

Acknowledgements This work was supported in part by the Natural Science Foundation of China under Grant 61703283, 61773328, 61672358, 61703169, 61573248, in part by the Research Grant of The Hong Kong Polytechnic University (Project Code:G-YBD9 and G-YBD9), in part by the China Postdoctoral Science Foundation under Project 2016M590812, Project 2017T100645 and Project 2017M612736, in part by the Guangdong Natural Science Foundation under Project 2017A030310067, Project with the title Rough Sets-Based Knowledge Discovery for Hybrid Labeled Data and Project with the title The Study on Knowledge Discovery and Uncertain Reasoning in Multi-Valued Decisions.

References

1. Ngan, Y., Pang, K., Yung, H.: Automated fabric defect detection—a review. Image Vis. Comput. **29**(7), 442–458 (2011)
2. Raheja, J., Kumar, S., Chaudhary, A.: Fabric defect detection based on GLCM and Gabor filter: a comparison. Optik Int. J. Light Electron Opt. **124**(23), 6469–6474 (2013)
3. Hu, G., Zhang, G., Wang, Q.: Automated defect detection in textured materials using wavelet-domain hidden markov models. Opt. Eng. **53**(9), 093107 (2014)
4. Wen, X., Cao, J., Liu, X., Ying, S.: Fabric defects detection using adaptive wavelets. Int. J. Clothing Sci. Technol. **26**(3), 202–211 (2014)
5. Tong, L., Wong, W., Wong, C.: Differential evolution-based optimal Gabor filter model for fabric inspection. Neurocomputing **173**, 1386–1401 (2016)
6. Ngan, Y., Pang, K., Yung, H.: Motif-based defect detection for patterned fabric. Pattern Recogn. **41**(6), 1878–1894 (2008)
7. Ngan, Y., Pang, K., Yung, H.: Ellipsoidal decision regions for motif-based patterned fabric defect detection. Pattern Recogn. **43**(6), 2132–2144 (2010)
8. Zhou, J., Semenovich, D., Sowmya, A., Wang, J.: Dictionary learning framework for fabric defect detection. J. Text. Inst. **105**(3), 223–234 (2014)
9. Li, P., Liang, J., Shen, X., Zhao, M., Sui, L.: Textile fabric defect detection based on low-rank representation. Multimedia Tools Appl. 1–26 (2017)
10. Tolba, A.: Neighborhood-preserving cross correlation for automated visual inspection of fine-structured textile fabrics. Text. Res. J. **81**(19), 2033–2042 (2011)
11. Çelik, H., Dülger, L., Topalbekiroğlu, M.: Development of a machine vision system: real-time fabric defect detection and classification with neural networks. J. Text. Inst. **105**(6), 575–585 (2014)

Robust Feature Extraction for Material Image Retrieval in Fashion Accessory Management

Yuyang Meng, Dongmei Mo, Xiaotang Guo, Yan Cui, Jiajun Wen
and Wai Keung Wong

Abstract Fashion accessory plays an important role in costume designing. A well-designed accessory consisting of different types of materials help enhance the aesthetic of the dresses. A key problem of accessory design is to find the replaceable material with appropriate aesthetic and cheaper price. However, such a process is performed manually in accessory factory, in which the work efficiency is very low. Therefore, material image retrieval is an important technique to automatic and facilitates the process of accessory design and management. In this paper, a voting-based preprocessing method is proposed to locate the material in the image. And thus a regression model is built to make use of the neighboring edge directions to optimize the robust edge direction of a point. Finally, both color and edge features will be coded as histogram-based features for representing the materials for image retrieval. Experiments have been conducted on real captured material image to validate the effectiveness of the proposed locating and searching technique.

Keywords Accessory · Material retrieval · Feature extraction

Y. Meng
School of Information Science and Engineering,
Shaoguan University, Shaoguan 512005, China

D. Mo · X. Guo · J. Wen (✉)
College of Computer Science and Software Engineering,
Shenzhen University, Shenzhen 518055, China
e-mail: enjoy_world@163.com

Y. Cui
School of Mathematics and Information Science,
Nanjing Normal University of Special Education, Nanjing 210038, China

J. Wen · W. K. Wong
Institute of Textile and Clothing, The Hong Kong Polytechnic University,
Kowloon, Hong Kong

J. Wen · W. K. Wong
The Hong Kong Polytechnic University
Shenzhen Research Institute, Shenzhen 518055, China

© Springer Nature Switzerland AG 2019
W. K. Wong (ed.), *Artificial Intelligence on Fashion and Textiles*,
Advances in Intelligent Systems and Computing 849,
https://doi.org/10.1007/978-3-319-99695-0_36

1   Introduction

Fashion accessory has made a great contribution to enhance the aestheticism of clothing. Generally speaking, a fashion designer needs to find a good combination of the
materials to construct an art of accessory (see Fig. 1), among which replacing the
expensive materials with the appropriate aesthetic and cheaper price is an important task. However, in reality it always takes the designer a long period of time to
accomplish such a process.

The image-based material retrieval is a promising technique that helps automate
and facilitate the task of accessory designing. It offers an intuitive way for the designer
to find the replaceable material that he wants. The difficulties of vision-based image
retrieval lie in the lighting condition, shape deformation and complexity of the image
background. Recently effective feature descriptors [1–3] have been designed for the
purpose of handling the above-mentioned problems, among which color and shape
are two important features for the retrieval task. In Chatzichristofis et al.'s work
[1], a two-stage fuzzy strategy was proposed to extract the color features from HSV
color space and thus the extracted features were further encoded as bag-of-visual-
words to increase the robustness of feature representation. Biswas et al. [2] held that
a number of distributed features computed from the interest points on the contour
are quite robust to describe the shape of the target. Similarly, Bhattacharjee et al.
[3] proposed to extract a series of feature points that can capture finer details of the
shape to resist the influence of rotation transformation under complex situations.
Moreover, even though good performance were reported by the newly proposed
features such as binary angular pattern [4], sketch descriptors [5], and hierarchical
string cuts [6], these methods take too much computation cost and cannot achieve
reliable performance in complex situation.

In this work, an automatic material locating technique and material retrieval technique will be developed to address the material searching problem. The first advantage of the proposed technique lies in the automatic scheme for material locating. The
second advantage of the proposed technique lies in the robustness of the extracted
edge direction feature, and the effective use of both color and direction feature.

The rest of the paper is organized as follows. Section 2 presents the proposed
method. Section 3 demonstrates the validation of the proposed method by the experiments. Finally, we conclude the paper in Sect. 4.

Fig. 1 Fashion accessory samples

2   The Proposed Method

In this work, a material database will be constructed for the research. Therefore, we first take pictures of the material samples to obtain material images first. An automatic material locating technique is needed to extract the material from the captured image. Thus we design a robust material retrieval technique by using joint color and direction feature. The details of the proposed method are presented below.

2.1   Automatic Material Locating

Due to the reflection problem of the material, we need to lay the material sample in the center of a special ring device with white color for image capturing, please see Fig. 2a, which helps reduce the reflection as much as possible on the captured image. Technically, we need to store the exact region of the material in the database but not the redundant region around such a material.

To this end, an automatic material locating method is developed. This method helps extract a single material from the whole image. Our idea is to check the neighbor points a and b of the j-th point on a contour of the captured image. For our target inner contour, majority of the values of point a must be 255, meanwhile majority of the values of point b must be 0. If the target inner contour is found, it is easy to extract the material. Thus we summarize the entire algorithm in Table 1.

2.2   Material Retrieval

Two notable features of the material are color and shape. Therefore, this work will perform color and shape analysis to extract both features. For color analysis, back

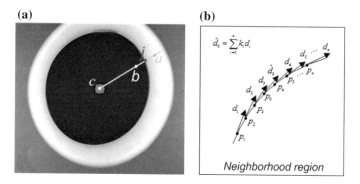

Fig. 2 Ideas of automatic material locating and robust edge direction estimation

Table 1 Algorithm of automatic material locating

Input: A captured material image, $t=0$, $w=0$, and $d=0$.
1. Transfer the image to gray image and smooth it with Gaussian filter.
2. Use dynamic threshold method to obtain the binary image.
2. Extract n (all the) contours on the binary image.
3. for $i=1,2, \dots ,n$ do
4.   Compute the center point c, and the i-th contour has m points.
5.   for $j=1,2, \dots ,m$ do
4.   Use point j and center c to compute the outer point a and inner point b.
5.   If the gray value of pont a is 255, thus $w=w+1$.
5.   If the gray value of pont b is 0, thus $d=d+1$.
6.   end for
7.   Normalize w and d. If $w+d>t$, and thus $t=w+d$, $k=i$.
5. end for
6. Segment the material in the specified region within the k-th contour.
Output: The extracted material.

projection technique will be applied to estimate the probability of the colors based on

$$\hat{q}_u = \sum_{i=1}^{n} \delta\left[c\left(x_i^*\right) - u\right] \tag{1}$$

where $m=64$ color bin was used, $u=1,2,\dots, m$ and function $c : \Re^2 \to \{1, 2, \dots, m\}$ transforms the i-th pixel value to the bin number u. Thus we normalize the probability by

$$\left\{ \hat{p}_u = \min\left(\frac{255}{\max(\hat{q})} \hat{q}_u, 255 \right) \right\}_{u=1,2,\dots,m} \tag{2}$$

to compute the corresponding probabilities $\hat{p}_u (u = 1, 2, \dots, m)$.

Due to the factors of image noises, inaccurate segmentation, and influence of complex background, for shape analysis, a regression model is designed to obtain the robust direction of the points along the contour of the shape. As shown in Fig. 2b, for a point P_0 along the contour, its direction $\hat{d}_0$ will be estimated by a series of neighbor points $P_1,\dots, P_n$ according to the following optimization model:

$$\arg \min_k \|d_0 - Dk\|^2$$
$$\text{s.t.} \|k\| = 1 \tag{3}$$

where $D = \begin{bmatrix} d_1 & d_2 & \cdots & d_n \end{bmatrix}$, and $k = \begin{bmatrix} k_1 & k_2 & \cdots & k_n \end{bmatrix}^{\mathrm{T}}$. Solving model (3) to obtain $\hat{d}_0 = \sum\limits_{i=1}^{n} k_i d_i$. Thus we divide the direction into eight bins and calculate the edge direction probability of the shape according to Eqs. (1) and (2). Finally, both color and direction features will be combined to represent the material in feature space. For a query material image, its combined color and direction feature will be computed and compared with the features of the existing material images in the database. The nearest distance in feature space will indicate the category of the query material image.

3 Experiments

3.1 Material Image Database

In this paper, a database with about 5000 material images was built. Each material has its own shape, color and size, which are useful properties for the recognition task. A few samples of the material are shown in Fig. 3.

3.2 Experiments on Automatic Material Locating

In this section, the materials were laid in the center of the ring device in order to capture the images with less reflection. And thus the proposed automatic material locating method will run on the captured images. The locating results are shown in Fig. 4 which demonstrates that the proposed method can automatically extract different kinds of materials perfectly under the said setting. It only took the proposed algorithm around 200 ms to finish the automatic locating task, which is quite efficient for the practical use.

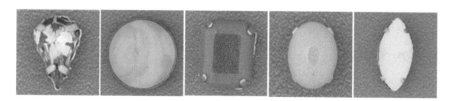

Fig. 3 Demonstration of material samples

Fig. 4 Automatic material locating results

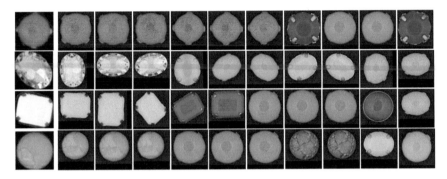

Fig. 5 Searching results of the materials. The first column shows the query images. The rest columns show the searching results

3.3   *Experiments on Material Retrieval*

As soon as the material database is well prepared, we performed the proposed retrieval technique for searching the similar materials as the designer wants. A visual demonstration of the proposed searching technique is shown in Fig. 5, which shows that the proposed technique not only can find the same material with different directions and reflections, but also can feedback similar materials for the designer's reference. The proposed technique is quite suitable to use in the fashion material searching task. It is because of the proposed robust edge direction model, the shape analysis and representation is pretty well.

4   Conclusion

In this paper, an automatic material locating technique is proposed for extracting the material from the captured image, which facilitates to build the material database. Moreover, we propose an edge direction estimation method for extracting the edge direction features robustly, which help resist the image noises, inaccurate segmentation and complex image background to represent the shape of the material perfectly. And thus both features will be represented and combined by using the probability idea. Experimental results validate the effectiveness of the proposed method. In the future, we will enlarge the scale of the database and design more robust and efficient algorithms to address the material searching problem to further facilitate the automation of accessory designing.

Acknowledgements This work was supported in part by the Natural Science Foundation of China under Grant 61703283, 61773328, 61672358, 61703169, 61573248, in part by the research grant of the Hong Kong Polytechnic University (Project Code: G-UA2B) in part by the China Postdoctoral Science Foundation under Project 2016M590812, Project 2017T100645 and Project 2017M612736, in part by the Guangdong Natural Science Foundation under Project 2017A030310067, Project with the title Rough Sets-Based Knowledge Discovery for Hybrid Labeled Data and Project with the title The Study on Knowledge Discovery and Uncertain Reasoning in Multi-Valued Decisions, and in part by the Shenzhen Municipal Science and Technology Innovation Council under Grant JCYJ20160429182058044.

References

1. Chatzichristofis, S.A., Iakovidou, C., Boutalis, Y., Marques, O., Co.Vi.Wo.: Color visual words based on non-predefined size codebooks. IEEE Trans. Cybern. **43**(1), 192–205 (2013)
2. Biswas, S., Aggarwal, G., Chellappa, R.: An efficient and robust algorithm for shape indexing and retrieval. IEEE Trans. Multimedia **12**(5), 372–385 (2010)
3. Bhattacharjee, S.D., Yuan, J., Tan, Y.-P., Duan, L.-Y.: Query-adaptive small object search using object proposals and shape-aware descriptors. IEEE Trans. Multimedia **18**(4), 726–737 (2016)
4. Hu, R.-X., Jia, W., Ling, H., Zhao, Y., Gui, J.: Angular pattern and binary angular pattern for shape retrieval. IEEE Trans. Image Process. **23**(3), 1118–1127 (2014)
5. Polsley, S., Ray, J., Hammond, T.: SketchSeeker: finding similar sketches. IEEE Trans. Hum. Mach. Syst. **47**(2), 194–205 (2017)
6. Wang, B., Gao, Y.: Hierarchical string cuts: a translation, rotation, scale, and mirror invariant descriptor for fast shape retrieval. IEEE Trans. Image Process. **23**(9), 4101–4111 (2014)

Woven Fabric Defect Detection Based on Convolutional Neural Network for Binary Classification

Can Gao, Jie Zhou, Wai Keung Wong and Tianyu Gao

Abstract Fabric defect detection plays an important role in the textile industry. However, this problem is very challenging because of the variability of texture and diversity of defect. In this paper, we investigate the problem of woven fabric defect detection using deep learning. A convolutional neural network with multi-convolution and max-pooling layers is proposed. Moreover, a high-quality database, which covers the common defects in woven fabric with solid color, is built. The experiments conducted on the database indicate that the proposed model could obtain the overall detection accuracy 96.52%, which shows the potential of the model in practical application.

Keywords Woven fabric defect · Deep learning · Convolutional neural network
Binary classification

1 Introduction

Woven fabrics are one of the most commonly used materials in our daily life. However, fabric defects are unavoidably generated during the process of weaving and finishing, and the income loss thus caused can be normally up to 45–65% [1]. Therefore, fabric inspection is a necessary and essential step to ensure the quality of the woven fabric. Traditionally, fabric inspection is carried out by trained human inspectors, but this non-automated manner suffers from the problems of reliability,

C. Gao (✉) · J. Zhou
College of Computer Science and Software Engineering,
Shenzhen University, Shenzhen 518060, People's Republic of China
e-mail: 2005gaocan@163.com

C. Gao · J. Zhou · W. K. Wong
Institute of Textiles and Clothing, The Hong Kong Polytechnic University,
Kowloon, Hong Kong

T. Gao
School of Minerals Processing and Bioengineering, Central South University,
Changsha 410083, People's Republic of China

© Springer Nature Switzerland AG 2019
W. K. Wong (ed.), *Artificial Intelligence on Fashion and Textiles*,
Advances in Intelligent Systems and Computing 849,
https://doi.org/10.1007/978-3-319-99695-0_37

accuracy as well as inspection speed. It is highly desired, in the textile industry, that an automatic technique can be utilized to realize the process of fabric inspection.

Fabric defect detection has been a popular research topic for many years and many vision-based approaches have been proposed to detect the defects of fabrics. These approaches can be roughly classified into non-motif and motif. Further, non-motif-based approaches can be categorized into four groups: statistical, structural, model-based, and spectral approaches [2]. The statistical approaches usually use first-order statistics, i.e., mean and variance, and second-order statistics, i.e., auto-correlation function and co-occurrence matrices, to represent textural features in texture discrimination. However, it is very challenging to discriminate subtle defects from the standard textile texture. The structural methods consider texture as a composition of textural primitives. Mostly, texture analysis is performed by extracting the texture elements and inferring their replacement rules. Unfortunately, these approaches are only effective in segmenting defects from the texture that is very regular. The model-based approaches represent the fabric images as a stochastic model. Similar to statistical approaches, the model-based method also does not perform effectively in detecting small defects. Finally, the high textile texture periodicity permits the usage of spectral features for fabric defect detection, when yarns are regarded as basic texture primitives. Among the four classes of methods, spectral methods [3] are most widely utilized in fabric defect detection, since textural features extracted in the frequency domain are less sensitive to both noise and intensity variation than that in the spatial domain. In general, the techniques involved in spectral methods are Fourier transform, wavelet transform, and Gabor transform. Although promising results are achieved, there still have two main drawbacks: (1) they are usually of high computational cost, which is vital to the real-time inspection system; (2) without automatic parameter recommendation, the empirical parameters cannot well-handle different situations of the defect detection problems.

Although some methods are proposed to deal with the problem of finding the defect within the fabric, some works are still needed to be done. First, a large-scale and high-quality benchmark data is required in fabric detection. Second, due to a large variability of possible textiles and highly varying fabric defects, a robust and parameter-free fabric detection algorithm is very desired, which is more suitable for industrial applications. To address these problems, we propose a convolutional neural network for fabric defect detection.

The rest of this paper is organized as follows. Section 2 presents the framework of the proposed convolutional neural network for fabric detection. Section 3 reports the experimental results. The conclusion and future work are given in Sect. 4.

Table 1 Architecture of the proposed convolutional neural network for fabric detection

Layer type	Output size	Number of parameters
Conv2D_1, (3, 3), 32	(126, 126, 32)	
Activation_1, Relu	(126, 126, 32)	320
MaxPooling2D_1, (2, 2)	(63, 63, 32)	
Conv2D_2, (3, 3), 32	(61, 61, 32)	
Activation_2, Relu	(61, 61, 32)	9248
MaxPooling2D_2, (2, 2)	(30, 30, 32)	
Conv2D_3, (3, 3), 64	(28, 28, 64)	
Activation_3, Relu	(28, 28, 64)	18,496
MaxPooling2D_3, (2, 2)	(14, 14, 64)	
Flatten_1	12,544	
Dense_1	64	802,880
Activation_4, Relu	64	
Dropout_1	64	
Dense_2	1	65
Activation_5, Sigmoid	1	

2 Fabric Defect Detection with Convolutional Neural Network

2.1 The Architecture of the Proposed Convolutional Neural Network

The neural network is an effective technique for machine learning and pattern recognition [4]. In recent years, a variety of deep learning models have been proposed, such as deep stacked auto-encoder, convolutional neural network, and deep belief network. Among these models, the convolutional neural network has attracted much attention and received great success in image classification, object detection, and segmentation [5]. In this paper, the problem of fabric detection is considered as a binary classification, namely determining a fabric image as normal and defect ones. A multi-layer convolution neural network is employed to extract the discriminant features and identify the defective fabric images. The architecture of the proposed convolutional neural network is shown in Table 1.

2.2   Convolution Layer

In the convolution layer, the original input image or feature map is convoluted with the filters to form the initial results. After adding the bias, the convolution results are fed as the input of the activation function to output the feature map. In the proposed convolutional neural network, three convolution layers are employed to extract different level discriminant features and each convolution layer can be expressed as

$$y_j^l = f\left(\sum_{i \in M_j} x_i^l * w_{ij} + \text{bias}_j^l \right), \tag{1}$$

where y_j^l is the output of neuron j in layer l, x_i^l is the input of number i in layer l, bias_j^l is the bias of neuron j in layer l, and w_{ij} convolution kernel, the symbols '*' and '$f(\cdot)$' denote the convolution operation and an activation function, respectively.

2.3   Pooling Layer

Convolution layer is usually followed by pooling layer. In the pooling layer, the feature map after activation function is downsampled into the smaller one, which is beneficial for simplifying the network architecture and reducing the scale of parameters. In the proposed convolutional neural network, the downsampling of max-pooling is used.

3   Experiments and Analysis

3.1   Data Set and Parameter Setting

In the experiments, we are mainly concerned with the woven fabrics. Up to now, there is no satisfactory benchmark data with enough high-quality samples to test algorithm. Herein, we collect many fabric images by ourselves to build a data set.

The data set includes 600 defect-free and 600 defect images from different woven fabrics with solid color. Each image is captured by a high-resolution CCD camera, and be resized and normalized into 128 * 128 pixels. Some defects after preprocessing are shown in Fig. 1.

In the experiment, the train, validation, and test sets are set to 50, 20, and 30% samples of the data set, respectively. For the proposed convolutional neural network, cross-entropy is selected as the loss function, the optimizer is set to Adam, and the accuracy index is adopted as measurement metric. To avoid over-fitting, we use some

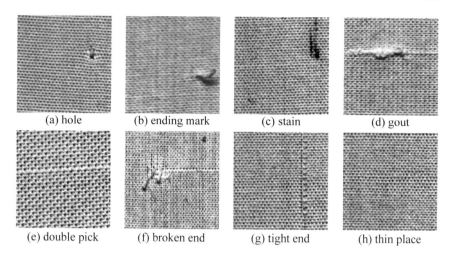

| (a) hole | (b) ending mark | (c) stain | (d) gout |
| (e) double pick | (f) broken end | (g) tight end | (h) thin place |

Fig. 1 Common defects in woven fabric

Table 2 Confusion matrix of the fabric detection algorithms

	Defective	Normal
Detected as defective	TP	FP
Detected as normal	FN	TN

strategies of data augmentation such slight rotation (lower than 15°), horizontal flip and vertical flip, thus the real training data are much more than 600 images.

3.2 Evaluation Criterion

The performance of fabric detection algorithms can be represented by a confusion matrix as Table 2.

In the matrix, the symbols TP and FN denote the real defect images are correctly detected as defective ones and misclassified as normal ones, respectively. Correspondingly, the symbols TN and FP denote the real normal images are correctly detected as normal ones and misclassified as defective ones, respectively. Based on the confusion matrix, several measures can be defined, such as detection accuracy, detection rate, false alarm rate, etc. In this paper, we use the overall detection accuracy to evaluate the performance of the fabric detection algorithm, namely, the value of the formula $(TP+TN)/(TP+FP+FN+TN)$.

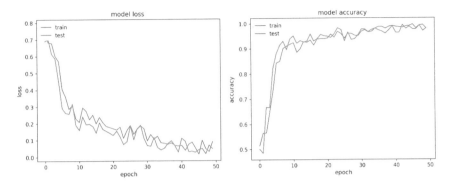

Fig. 2 The performance of the proposed convolutional neural network (epoch = 50)

3.3 Experimental Results and Analysis

In the experiments, we set the number of epochs as 50. The experimental results are shown in Fig. 2.

From Fig. 2, we can see that the proposed model almost fits the train and validation data after only 10 epochs and obtains over 90% detection accuracy with 0.2 overall loss. After 30 epochs, the model tends to converge and achieve 97.75% overall detection accuracy with only 0.1 overall loss. Although the model still performs until to 50 epochs, the performance is not much improved and terminated at the overall detection accuracy 98.49% with only 0.08 overall loss.

On the test set, the trained model is used to further evaluate the performance. The overall detection accuracy is 96.52%. After observing the misclassified fabric images, we find them can be roughly categorized to three cases: (1) normal image with blur; (2) defective image with slight thin/thick place; (3) defective image with a small defect at the border. Some examples are shown in Fig. 3. The fabric image with motion blur could mislead the model and be wrongly recognized as the defective one. The fabric with slight thin/thick place is difficult to discriminate from the normal one, even human inspector can make this mistake, therefore, the model fails to detect this type of defects. As for the third case, the model uses the three convolution layers and some features in the border are thus removed. Border filling or defect region centralization may alleviate this problem.

4 Conclusion

In this paper, we propose a convolutional neural network for fabric detection. The detection of fabric defects is considered as the problem of binary classification, and the neural network with three convolution layers are designed to extract the discriminated features. Moreover, a database that covers the common defects in the

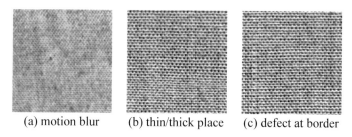

(a) motion blur   (b) thin/thick place   (c) defect at border

Fig. 3 The misclassified fabric images

woven fabric is built. The experiment results show that the proposed convolutional neural network could attain the satisfactory detection accuracy. In the future, we try to collect more fabric images with the defects and design a better neural network to improve the detection accuracy.

Acknowledgements This research is supported by the General Research Fund of the Research Grant Council, Hong Kong (Project No.: 15202217), the Guangdong Natural Science Foundation (Project No.:2018A030310451, 2018A030310450).

References

1. Srinivasan, K., Dastoor, P.H., Radhakrishnaiah, P., Jayaraman, S.: FDAS: A knowledge-based framework for analysis of defects in woven textile structures. J. Text. Inst. **83**, 431–448 (1992)
2. Ngan, H.Y.T., Pang, G.K.H., Yung, N.H.C.: Automated fabric defect detection-A review. Image Vis. Comput. **29**(7), 442–458 (2011)
3. Kumar, A., Pang, G.K.H.: Defect detection in textured materials using Gabor filters. IEEE Trans. Ind. Appl. **38**, 425–440 (2002)
4. Li, Y., Zhao, W., Pan, J.: Deformable patterned fabric defect detection with fisher criterion-based deep learning. IEEE Trans. Autom. Sci. Eng. **14**(2), 1256–1264 (2017)
5. LeCun, Y., Bengio, Y., Hinton, G.: Deep learning. Nature **521**, 436–444 (2015)

Complex Textile Products and Reducing Consumer Waste

Colin Gale

Abstract This paper discusses how AI, big data and new textile technologies could provide an integrated solution to reducing the environmental impact of textiles. Together material culture and consumerism have stimulated the kind of ecological impact that has given rise to the definition of the Anthropocene, that point at which our collective behavior has become a force of nature that is shaping the earth. We already acknowledge the environmental impact of fashion and textiles and societies have sought to limit the negative effects through environmental legislation and ethical appeals. However consumerism is universally an element of national economic strategies and personal consumption is, as yet, seen as both economic stimulus and an indicator of social prosperity and well-being. Consequently the mechanisms we have developed for constraining textile waste and pollution are perpetually in opposition to national political and economic agendas as well as the aspirations of individuals. In this complex and dystopian situation AI, big data and new textile technologies may offer a third way for the regulation of fashion and textile consumption. One that offers a set of interventionist strategies and creative opportunities that could alter consumer behaviors and expectations and consequently impact on both textile product lifecycles and retail models.

Keywords Anthropocene · Complexity · Consumerism · Products · Waste

1 Introduction

Thoughts about the impact of our species on the planet gave rise to a surge of political activism during the latter half of the twentieth century. In turn this stimulated reflection on the respective negative impact of textiles and garment production on the environment. Post Millenium we have a range of popular strategies that have been enthusiastically touted by many in academia and in industry as ways forward

---

C. Gale (✉)
School of Fashion and Textiles, Birmingham City University, Birmingham, UK
e-mail: colin.gale@bcu.ac.uk

© Springer Nature Switzerland AG 2019   315
W. K. Wong (ed.), *Artificial Intelligence on Fashion and Textiles*,
Advances in Intelligent Systems and Computing 849,
https://doi.org/10.1007/978-3-319-99695-0_38

for example eco textiles, vintage, recycling, slow fashion, even luxury. However the experience of any modern city dweller is that there is a glut, an oversupply of garments and other textile products. This is borne out by a number of indicators—overstocking, heavy discounting, and the failure of retail chains. The fast fashion model has become so powerful that it has created an increasingly unsustainable market environment coupled with a throwaway culture. On reflection the various strategies to reduce the impact of textile and garment production do not, in and of themselves, actually significantly impede growth in the volume of production. As the earth's population relocate and move to cities and as per capita wealth increases we are unlikely to halt the trend of increasing fashion and textile production without finding another solution. The clues lie in our habits, why we keep some things and why we throw others away.

2 Sustainability Strategies

Existing strategies for reducing the environmental impact of fashion and textiles fall into a few categories—reducing the chemical footprint of textile production, extending the product lifecycle and utilizing value and sentiment. Indications are that existing strategies for reducing impact have little effect with regard to reducing production, in fact production, with all its consequences, is set to increase [1] not least due to potential demand in emerging markets. If we were looking to identify the key objectives of a sustainable global fashion and textile industry it would have to include reducing consumption but this leads to the counter-intuitive ethical proposition of advanced consumer economies in that we want to sell more but throw away less. People need to buy less clothes but retailers need to find another way to make money from the clothes they sell. The solutions to what appears an impossibility are to be found within other product sectors [2] and fashion and textiles needs to change its product offer accordingly.

3 The Phone, the Car and the Coat

There are various factors that cause consumers to hold onto some products much longer than others, such as scarcity, need, dependency, and price. At one end of the spectrum is the mobile phone, most of us know the ensuing panic if we lose or forget our phone and yet the dependency is not intrinsic to the object itself. Neither is our dependency of a single dimension, it is a need to be connected, a need for access to information and more recently a genuine addiction to social stimulus [3]. Cost is a factor but increasingly the market has made that less painful, however the majority of us for purposes of the economic arrangement, stay with just the one phone until its damaged or it becomes (in terms of the design culture of the phone including its functionality) old. Most of us have one phone that we use most of the time, we feel we cannot be without it and (most of us) rarely change it that often. Our dependency on the car is different. It is one of the major purchases we can make, it depreciates significantly the minute we buy it, consequently we feel we must maximize our usage to 'get value'. We have various notions of need about the car relating to personal or family transportation, shopping, holidays and so on. Again most of us have one car that we use often sometimes daily, we feel we cannot be without it and (most of us) rarely change it that often. The average Briton owns fourteen cars in their life [4] and the majority of households in the UK (45%) have only one car.

The phone and the car are different types of object yet from a personal perspective, in relation to both and for different reasons, we will not own so many in our lives, we make daily use of them and replacing them can be a reluctant decision. If we compare these products with a coat, some of us probably have a few, most of us it seems may have a lot. Apart from functional clothing that serves a specific purpose, we have no substantive dependency on any coat we own, they are in most instances interchangeable. Many of us also like buying new coats at regular intervals. In general the behavior pattern of consumers in relation to clothing has increasingly become one of hoarding [5]. We have wardrobes full of clothes and yet we carry on buying.

The fashion industry would no doubt express a type of essentialism about consumption, that we need style change and choice, we must have new to keep up. The sociologists and psychologists might say our need to keep buying is about identity and our need for self-expression. The reality is we probably buy clothes because they are increasingly cheap, easy to get, we enjoy the shopping experience, they are easy to store and we have no specific need for any one item, nor are dependent on it. At present there is no real indication that this global consumer mindset or pattern of behavior is going to change.

4   Moving on from the Era of Box-Shaped Appliances

Since the 1980s computing technology and all its adjuncts in material product terms have followed a few paths—miniaturization, relatively anonymous box-shaped appliances with buttons, a visual window of simulations and yet more (virtual) buttons. Overridingly the connection with our hands has determined the shape of the technology. The thesis of the Internet of Things, the world of smart interconnected objects, actuators, and switches, while recognizing the holism of objects with communication functions is however not a manifesto to make all of our material cultural objects communicative or connected. This world of boxes, buttons, switches, and robots is thus still very far from that of a textile, there is in a sense no essential need to make textiles smart and we accept and almost expect our material culture to be discontinuous, i.e., some things will not be of the Internet of Things. Consequently most technologists and scientists would see smart textiles as niche and peripheral, equally most fashion designers and consumers seek and expect no change in the computing and communication properties of textile goods and garments although recent innovations suggest such changes are far more possible than a decade ago [6].

When we consider the profligacy or our fashion and textile consumption and compare it with our more moderate purchasing of phones and cars, it is a reasonable supposition that if we had the same degree of dependency on a single clothing item as we do with our phones (and the item itself lasted long enough) we would be inclined to wear it all the time. At present no such clothing items exist. Let us suppose for sake of argument that such a clothing item existed and to all intents and purposes replaced the functionality and interactive features of your phone. In what senses might it not be successful? Objections to the viability of such a garment might include invariant style, hygiene, performance requirements, durability, changes in sizing and so on. If such objections could be overcome single items of clothing would have a different life, not only would we wear them a lot, they might also be passed on to others. In undertaking this hypothetical argument incidentally we have identified the technological goals for future smart clothing and the altered mindset that the consumers, producers, and retailers need to pursue.

Earlier attempts to integrate computing technologies with textiles, for reasons of topologies and material properties, made little progress and have given rise to the technical solution of discrete technology on or in textiles using techniques like pocketing, for example Sensoria's monitoring technologies [7]. At the same time the aforementioned drive of technology itself toward miniaturization facilitated these approaches. The early ambitions in relation to smart textiles had perhaps the hope of mimicking the printed circuit board (PCB) [8] and components such as processors. Currently flexible and cheap substrates such as foils and paper can be used in conjunction with printed electronics to produce lightweight and flexible component technology [9]. The way in which textiles are normally manufactured still precludes the integral production of a computing device as a garment but progress has been made on techniques for incorporating electronic devices and recent manufacturing technologies provide us an opportunity to revisit the topological complexities of integration through 3D printing or complex knitting [10]. The challenge of creating smart textiles and garments has a longer development arc than that for the flat hard object but it seems recent progress begins to address the technological goals for smart clothing.

5   Changing Clothing Culture

Of the goals identified for achieving durable clothing with high consumer dependency, none is more key in fashion design and retail than the language of difference, for example differences in shape, texture, and surface decoration. If new forms of garment have transformative properties then it begins to address the need for expressive and artistic scope and perhaps even offers new channels for communication [11]. Another issue is the ability of wearers to be able to control garment change or perhaps autonomous garment change utilizing personalized data. It seems commonsensical that such garments would not mimic the clothes we know, that adoption of a new form of clothing would require movement away from traditional clothing.

The drivers for adopting such clothing might be style and trend but the key driver to achieve an extended product usage will be computing and communication technology, in essence the functions of our phones need to be absorbed by what we wear [12]. Many of the transformative and communication capacities we would seek in our future clothing we already rehearse as a society in our science fiction and fantasy movies, the next step will be to make our imaginings tangible and real.

One of the greatest challenges will be how we can extend product durability and hence lifecycle. We are used to a tradition of the soft and comfortable, consequently the idea of durable garments seems antithetical to the character of textiles. Our attempts so far at durable synthetics have created consternation and damage to the environment and life [13]. Presently durable gadgets provide the portability of communications and the expectation of smart clothing is that it would continue to be wearable and by default, disposable and lacking durability. However the difficult problem of durability is not an excuse for putting off the introduction of product dependency as a stratagem for reducing clothing and textile waste.

In the era of the Anthropocene we have become conscious of our impact on the world and yet our attitude towards our material and designed culture carries on as if each product area is independent, as if our material culture is not interconnected, that it will always be seen as a restless diversity and not a singularity for which we are all responsible. However these attitudes are not sustainable and they belong to a time before global consciousness about the environment. If we and the planet are to survive the consequences of mass consumerism then our products must either have the properties of the permanent or the ephemeral. Fashion and textiles are currently very far from these qualities however the constituent technology and pattern of human behavior required to change the fundamental nature of what we wear and why we wear it is, at least in part, present. This should be sufficient to encourage us.

6   Conclusion

On reflection for at least 20 years, since the turn of the Millenium, various solutions have been proposed to reduce the environmental impact of textile and garment production. In essence, whether aimed at the supply chain, the consumer or post-consumer product lifecycle, these have all been strategies for damage limitation, for example: shifting crop types and fiber sources to less water or insecticide dependent species; changing post-processing; politicizing the consumer or the designer. However none of these strategies can really be said to have reduced the quantity of production and indications are that as new industrial and post-industrial societies emerge, production will actually significantly increase. It is unrealistic to expect that consumers will become so radicalized they will volunteer to buy less or alternately that governments will become so conscientious they seek to halt a major sector of their circular economy. Consequently there is only one alternative solution, that our relationship with what we wear changes and that is only likely to happen if the clothes themselves have a different nature and purpose. The ambition to produce smart clothing incorporating computing functionality is far from new but in many senses the technology which might underpin it is still evolving. Much of that technology has begun to resolve itself in other product arenas such as mobile phones and this in turn has evidenced how human expectations of objects and the nature of human–object interaction are transformed when technological functionality is available. In essence technology will allow clothing to be more useful and in so doing will increase our dependency on bespoke garments. If in turn we are capable of addressing other key factors such as durability and the ability of clothes to change appearance then we might genuinely alter the current pattern of fashion consumption and move to a time when each of us has no need or desire for say more than ten to twenty outfits in our lifetime. At present this seems utterly unbelievable, which in and of itself, shows, sadly, how locked we are in our present consumer mindset.

References

1. Statista.: Apparel market size projections from 2012 to 2025, by region (in billion U.S. dollars). https://www.statista.com/statistics/279757/apparel-market-size-projections-by-region/. Accessed 29 Apr 2018 (2018)
2. Gustavsson, V., Khan, M.: Monetising blogs: enterprising behavior, co-creation of opportunities and social media entrepreneurship. J. Bus. Ventur. Insights **7**, 26–31 (2017). https://doi.org/10.1016/j.bvi.2017.01.002
3. Huang, H.: Social Media Generation in Urban China: A Study of Social Media Use and Addiction among Adolescents (Understanding China). Springer, Berlin (2016). https://doi.org/10.1007/978-3-642-45441-7
4. Hull, R.: Can you Guess How Much We Each Spend on Motoring In A Lifetime. http://www.thisismoney.co.uk/money/cars/article-3991208/Can-guess-spend-lifetime-motorist.html. Accessed 29 Apr 2018 (2016)
5. Joung, H.: Materialism and clothing post-purchase behaviours. J. Consum. Mark. **30**(6), 530–537 (2013). https://doi.org/10.1108/jcm-08-2013-0666
6. Sawh, M.: The Best Smart Clothing: From Biometric Shirts to Contactless Payment Jackets. https://www.wareable.com/smart-clothing/best-smart-clothing. Accessed 29 Apr 2018 (2018)
7. Sensoria: http://www.sensoriafitness.com/platform/. Accessed 29 Apr 2018 (2018)
8. Post, E., Orth, M.: Smart Fabric, or "Wearable Clothing". http://web.media.mit.edu/~rehmi/pdf/00629937.pdf. Accessed 29 Apr 2018 (1997)
9. Nisato, G., Lupo, D., Ganz, S.: Organic and Printed Electronics: Fundamentals and Applications. Pan Stanford, Boulevard (2016)
10. Narayan, V., Albaugh, L., Hodgins, J., Coros, S., McCann, J.: Automatic machine knitting of 3D meshes. ACM Trans. Graph. 1(1). https://doi.org/10.1145/3186265 https://drive.google.com/file/d/1UO0aGgbZqidvzgupqrJ (2018)
11. Sensoree.: http://sensoree.com/artifacts/ger-mood-sweater. Accessed 29 Apr 18 (2018)
12. Kennemer, Q.: Your Shirt Could have a Flexible OLED Display Embedded in it in the Future. https://phandroid.com/2016/11/29/oled-display-fabric/. Accessed 29 Apr 2018 (2016)
13. Messinger, L.: How your Clothes Are Poisoning our Oceans and Food Supply. https://www.theguardian.com/environment/2016/jun/20/microfibers-plastic-pollution-oceans-patagonia-synthetic-clothes-microbeads. Accessed 29 Apr 2018 (2016)

A Fast Parallel and Multi-population Framework with Single-Objective Guide for Many-Objective Optimization

Haitao Liu, Weiwei Le and Zhaoxia Guo

Abstract A large number of objectives pose challenges to many-objective evolutionary algorithms (MOEAs) in terms of diversity, convergence, and complexity. However, the majority of MOEAs are not able to perform well in all three aspects at the same time. To tackle this issue, this paper proposes a fast parallel and multi-population framework with single-objective guide for many-objective optimization. The general framework is able to enhance diversity via sub-populations and maintain convergence by the information sharing between sub-populations. The proposed framework is implemented on three representative MOEAs and is compared with original MOEAs on 64 many-objective benchmark problems. Experimental results show that the proposed framework is capable of enhancing the performance of original MOEAs with satisfactory convergence, diversity, and complexity.

Keywords Multi-population framework · Many-objective optimization
Evolutionary algorithms · Single-objective guide

1 Introduction

Many real-world problems can be considered as multiobjective optimization problems (MOPs). Generally a MOP can be formulated as follows:

$$\min \mathbf{F}(\mathbf{x}) = (f_1(\mathbf{x}), \ldots, f_m(\mathbf{x}))$$
$$subject \text{ to } \mathbf{x} \in \Omega, \tag{1}$$

where $\Omega \subseteq \mathbb{R}^n$ is the n-dimensional decision space and $\mathbf{F}$ is a m-dimensional objective space. $\mathbf{x} = (x_1, \ldots, x_n)^T \in \Omega$ is a candidate solution and $\mathbf{F}(\mathbf{x})$ is the corresponding objective vector . $\mathbf{x}^2$ is dominated by $\mathbf{x}^1$ if and only if $f_i(\mathbf{x}^2) \geq f_i(\mathbf{x}^1)$ for

H. Liu · W. Le · Z. Guo (✉)
Business School, Sichuan University, Chengdu 610065, People's Republic of China
e-mail: zx.guo@alumni.polyu.edu.hk

© Springer Nature Switzerland AG 2019
W. K. Wong (ed.), *Artificial Intelligence on Fashion and Textiles*,
Advances in Intelligent Systems and Computing 849,
https://doi.org/10.1007/978-3-319-99695-0_39

all $i \in \{1, \ldots, m\}$ and $f_i(\mathbf{x}^2) \neq f_i(\mathbf{x}^1)$. If no other solution dominates a solution $\mathbf{x}^*$, $\mathbf{x}^*$ is a Pareto-optimal solution. All Pareto-optimal solutions are the member of the Pareto-optimal set (PS) and the corresponding objective vectors are called the Pareto front (PF). If the number of objectives is more than three, MOPs is often called as many-objective optimization problems (MaOPs). Generally, the aim in tackling MOPs is to obtain a number of uniformly distributed Pareto-optimal solutions along the PF.

Numerous multiobjective optimization evolutionary algorithms (MOEAs) have been developed to approximate the PF since 1990s [1, 2]. However, the performance of most MOEAs relying on Pareto dominance deteriorates drastically with the increase in the number of objectives. To this end, a number of many-objective evolutionary algorithms (MaOEAs) have been proposed and they focus on enhancing population diversity. Among them, decomposition- and indicator-based approaches have gained more popularity in solving MaOPs. To be specific, MOEA/D [3] decomposes original MaOP into a number of single-objective optimization problem by aggregating objectives with weight vectors. The diversity is increased by constructing uniformly distributed weight vectors. Compared to MOEA/D, NSGA-III [4] constructs uniformly distributed reference lines to enhance diversity. Among indicator-based approaches, AR-MOEA [5] shows promising versatility to tackle MOPs and MaOPs with different types of PF because it adaptively updates reference points, where the diversity is also maintained by reference points.

Although current MaOEAs show good performance on solving MaOPs, the computational cost of some MOEAs is prohibitively expensive, especially the indicator-based approaches. Therefore it poses a huge challenge for MaOEAs to solve real-world MaOPs because only one single fitness evaluation is time-consuming. To tackle this challenge without lessening diversity, this paper proposes a fast parallel and multi-population framework with single-objective guide for many-objective optimization, termed pMP/SG. It is well-known that the optimal solution to the optimization problem with a certain objective is close to or even lie in the PF. Thus, we can use a single-objective process, which generates the solutions to corresponding single-objective optimization problems, to accelerate the convergence to the PF. Besides, multiple populations can enhance diversity without additional computational cost, and reduce complexity of algorithms. Parallel processing can also speed up the execution times.

This paper contributes to MaOEAs literature from two perspectives. First, we propose a fast parallel and multi-population framework with single-objective guide for many-objective optimization. Second, a novel information sharing mechanism between sub-populations is proposed.

2 Proposed Algorithm

2.1 Algorithm Overview

The basic framework of the pMP/SG is presented in Algorithm 1. First, the initial population P with N individuals is randomly generated. Based on initial population P, the three stages are repeated until the stopping criterion is satisfied. Stage 1 is a multi-population single-objective guide process. In this stage, the initial population P is divided randomly up M sub-populations with the same population size, where M is the number of objectives. By taking sub-population P_m as the initial sub-population, the best solutions to the single-objective optimization problem with the ith objective of the investigated MaOP is then found by I_1 iterations. After that, the M evolved sub-populations are merged into a new population P. It is worth noting that we can choose the single-objective optimization approaches according to the property of MaOPs. In stage 1, the individuals would not be influenced by conflicting objectives and the sub-populations can be guided by the corresponding objective to search different regions of the objective space. Stage 2 is a multi-population many-objective optimization process. In this stage, the investigated MaOP is optimized in parallel by using M sub-populations that are generated by randomly dividing population P into M sub-populations with the same population size. For each sub-population P_m, we define an export set P_m^e and assign all individuals of sub-population P_m to the export set. To balance diversity and convergence, information on different sub-populations should be shared when a certain condition, such as a constant generation I_2 is satisfied. Each sub-population chooses its partner and selects individuals from the sub-population itself and its partner. Each new sub-population optimizes all objectives of a MaOP based on a MOEA by I_2 iterations in parallel and then updates its export set. The procedure of the information sharing and parallel evolution is performed F times. After that, the M evolved sub-populations are merged into a new population P again. The information sharing mechanism is introduced in detail in Sect. 2.2. Stage 3 is a single-population many-objective optimization process. To achieve better convergence, the merged population P optimizes the MaOP by I_3 iterations. Finally, the final population P is output.

The implementation details of main components of pMP/SG are presented in the following subsections.

Algorithm 1: General Framework of pMP/SG
Input: N(population size), M(the number of objectives), I_n(iterations of stage$n$).
Output: P(final population)

1 Randomly generate initial population P;
2 **while** *the stopping criterion is not met* **do**

/*Stage 1: Multi-population single-objective optimization*/
3 Divide randomly P into M sub-populations with the same population size
   $P_1, \ldots, P_M$;
4 **for** $m=1$ to M **do**
5     $P_m \leftarrow$ Evolution(P_m, m, I_1)/* optimizing the mth objective */
6 **end**
7 $P = \cup_{m=1}^{M} P_m$;
   /*Stage 2: Multi-population many-objective optimization*/
8 Divide randomly P into M sub-populations with the same population size
   $P_1, \ldots, P_M$;
9 **for** $f = 1$ to F **do**
10    $P_1, \ldots, P_M \leftarrow$ Information Sharing($P_1, \ldots, P_M$)
11   **for** $m = 1$ to M **do**
12       $P_m \leftarrow$ Evolution(P_m, I_2)/* optimizing all objectives */
13   Update export set: $P_m^e \leftarrow P_m$;
14   **end**
15    $f \leftarrow f + 1$;
16 **end**
17 $P = \cup_{m=1}^{M} P_m$;
   /*Stage 3: Single-population many-objective optimization*/
18 $P \leftarrow$ Evolution(P, I_3)/* optimizing all objectives */
19 Return P.

2.2 Information Sharing Mechanism Between Sub-populations

To improve diversity and accelerate convergence, we propose a novel information sharing mechanism. Particularly, each sub-population only selects one another sub-population as a partner for information sharing, and each sub-population can be chosen as a partner for only once. That is to say, each sub-population can be chosen as a partner for another sub-population and participate in the procedure of information sharing.

To this end, we first generate a random permutation ψ of 1 to M and update the permutation by adding the first element of ψ at the end of ψ. This is able to make fully use of the valuable information of each sub-population, and maintain population diversity by keeping unique individuals as much as possible. The sub-population is updated by selecting individuals from current sub-population and export set of its partner via the selection operator used in GrEA [6].

3 Experimental Studies

In order to validate the effectiveness of pMP/SG framework, the pMP/SG framework is implemented based on three representative MOEAs, namely AR-MOEA [5], NSGA-III [4] and MOEA/D [3]. We compare the resulted three algorithms, denoted by pAR-MOEA, pNSGA-III, and pMOEA/D, with original AR-MOEA, NSGA-III and MOEAD, respectively. Experimental settings and comparative results are presented in the following sections.

3.1 Experimental Settings

We employ 64 problem instances from three widely used test suites, namely DTLZ1-DTLZ7 [7] and WFG1-WFG9 [8]. The population size is set to 300, 400, 400, and 600 for $M \in \{3, 8, 10, 15\}$. The maximum iteration is set to 250. For pMP/SG, The generations (I_1, I_3) in stage 1 and stage 3 are set to 300 and 50 respectively. The interval I_2 and frequency F of information sharing in stage 2 are set to 50 and 4 respectively. Original genetic algorithm [9] is used for single-objective optimization in stage 1. All algorithms are implemented in PlatEMO [10] without modifying other parameters.

The inverted generational distance (IGD) [5] is employed to evaluate the performance of algorithms. To calculate IGD, roughly 10,000 evenly distributed points are sampled on the PF via Das and Dennis's method [11]. Besides, each algorithm is independently run 20 times for each instance. The Mann–Whitney–Wilcoxon ranksum test with a significance level of 5% is employed to perform statistical analysis. Symbols '+', '−' and '≈' indicate that the result obtained by proposed framework is significantly better, worse and similar to that obtained by compared MOEAs.

4 Results

Table 1 presents the mean of IGD values obtained by pNSGA-III and NSGA-III on 64 MaOP instances, including DTLZ1-DTLZ7 and WFG1-WFG9 problems with 3-, 8-, 10-, and 15-objective. It can be seen that pNSGA-III has obtained significantly better results on 46 of 64 MaOP instances, whereas NSGA-III only obtained significantly better results on 6 instances. Similarly, both pAR-MOEA and pMOEA/D obtained significantly better results on 28 of 64 MaOP instances, and worse results on 18 and 17 instances, respectively. Hence, the pMP/SG framework is able to improve the performance of existing MOEAs.

Apart from better optimization performance, pMP/SG framework is also able to lower the computational cost of some MOEAs. For example, the computation complexity of AR-MOEA is $O(MN^3)$. The computation complexity of stage 2 of

Table 1 Mean of IGD results obtained by pNSGA-III and NSGA-III on DTLZ1-DTLZ7 and WFG1-WFG9 problems with 3-, 8-, 10-, and 15-objective

Obj.	$M = 3$		$M = 8$		$M = 10$		$M = 15$	
Problems	pNSGA-m	NSGA-III	pNSGA-III	NSGA-III	pNSGA-III	NSGA-III	pNSGA-III	NSGA-III
DTLZ1	0.01095 −	0.01082	0.11197 ≈	0.10020	0.17974 ≈	0.20736	0.16937 ≈	0.20865
DTLZ2	0.02885 +	0.02858	0.29520 +	0.29834	0.40809 +	0.49461	0.49613 +	0.61414
DTLZ3	0.02915 +	0.04286	1.65166 +	3.10794	4.23454 +	12.50985	7.86648 +	13.24657
DTLZ4	0.05452 ≈	0.02859	0.33555 −	0.30230	0.44098 ≈	0.44534	0.51183 +	0.56070
DTLZ5	0.00400 ≈	0.00378	0.23254 +	0.47317	0.23246 +	0.63392	0.19960 +	0.36042
DTLZ6	0.00602 ≈	0.00614	0.50277 +	0.93907	0.40986 +	4.20541	0.27523 +	3.03499
DTLZ7	0.03830 ≈	0.03852	0.60642 +	0.69129	1.03444 +	1.31386	4.11292 +	6.52105
WFG1	0.13454 +	0.38731	0.89692 +	1.49068	1.23456 +	1.87076	1.77192 +	2.51100
WFG2	0.10988 ≈	0.11078	1.20403 ≈	1.77480	2.35414 +	5.06475	4.18866	10.03037
WFG3	0.04212 +	0.06526	0.80508 ≈	0.98103	0.55163 +	0.92567	1.39715 +	2.14667
WFG4	0.11816 +	0.11911	2.58589 +	2.64557	3.78188 +	4.60801	7.18241 +	8.19685
WFG5	0.13919 −	0.13806	2.60292 +	2.63627	4.20591 +	4.47525	6.89640 +	7.93109
WFG6	0.14235 +	0.15931	2.61570 +	2.67736	4.30907 +	4.68507	7.37676 +	8.70488
WFG7	0.11951 −	0.11807	2.59260 +	2.65818	4.25386 +	4.55577	6.99014 +	8.27268
WFG8	0.22767 −	0.21852	2.77435 ≈	2.98949	4.73342 +	5.12317	8.21691 +	8.66245
WFG9	0.22767 −	0.21852	2.55500 +	2.60009	4.00879 +	4.36951	7.23003 +	8.03777
+/−/≈	46/6/12							

pMP/SG is $O\left(N^3/M^2\right)$. Therefore, with the increase in the number of objectives, pMP/SG framework shows significant superiority in terms of computational cost. This observation is also supported by experimental comparisons.

5 Conclusion

This paper proposes a fast parallel and multi-population framework with single-objective guide for many-objective optimization, called pMP/SG. The pMP/SG framework is implemented based on three representative MOEAs. Empirical results on 64 benchmark problems demonstrated that the pMP/SG is a general and effective framework in developing versatile MOEAs for solving MaOPs. Furthermore, the novel information sharing mechanism between sub-populations is able to balance diversity and convergence. The pMP/SG can also reduce the complexity of algorithms and lower the computational cost. In future, self-adaptive settings of algorithm parameters, such as dynamic sub-population size, and their effects on the many-objective optimization performance could be investigated.

References

1. Deb, K.: Current trends in evolutionary multi-objective optimization. Int. J. Simul. Multi. Design Optim. **1**(1), 1–8 (2007)
2. Jones, D.F., Mirrazavi, S.K., Tamiz, M.: Multi-objective meta-heuristics: an overview of the current state-of-the-art. Eur. J. Oper. Res. **137**(1), 1–9 (2002)
3. Zhang, Q., Li, H.: MOEA/D: a multiobjective evolutionary algorithm based on decomposition. IEEE Trans. Evol. Comput. **11**(6), 712–731 (2007)
4. Deb, K., Jain, H.: An evolutionary many-objective optimization algorithm using reference-point-based nondominated sorting approach, part I: solving problems with box constraints. IEEE Trans. Evol. Comput. **18**(4), 577–601 (2014)
5. Tian, Y., et al.: An indicator based multi-objective evolutionary algorithm with reference point adaptation for better versatility. IEEE Trans. Evol. Comput. **22**(4), 609–622 (2017)
6. Yang, S., et al.: A grid-based evolutionary algorithm for many-objective optimization. IEEE Trans. Evol. Comput. **17**(5), 721–736 (2013)
7. Deb, K., et al.: Scalable test problems for evolutionary multiobjective optimization. In: Evolutionary Multiobjective Optimization. Theoretical Advances and Applications, pp. 105–145 (2005)
8. Huband, S., et al.: A scalable multi-objective test problem toolkit. In: International Conference on Evolutionary Multi-Criterion Optimization. Springer, Berlin (2005)
9. Goldberg, D.E.: Genetic Algorithms in Search, Optimization and Machine Learning, pp. 2104–2116. Addison-Wesley Pub. Co, Boston (1989)
10. Tian, Y., et al.: PlatEMO: A MATLAB platform for evolutionary multi-objective optimization. IEEE Comput. Intell. Mag. **12**(4), 73–87 (2017)
11. Das, I., Dennis, J.E.: Normal-boundary intersection: a new method for generating the Pareto surface in nonlinear multicriteria optimization problems. SIAM J. Optim. **8**(3), 631–657 (1998)

Multiple Criteria Group Decision-Making Based on Hesitant Fuzzy Linguistic Consensus Model for Fashion Sales Forecasting

Ming Tang and Huchang Liao [ORCID]

Abstract In many real-world multiple criteria group decision-making process, people cannot provide accurate preference information over a set of alternatives because of the increasingly complex environment. Fashion sales forecasting can be taken as a multi-criteria group decision-making problem given that people need to consider product life cycle, year-on-year growth rate, seasonal factor, industry factor, and consumer factor comprehensively when they forecast the fashion scales. In this paper, we developed a fuzzy linguistic model for fashion sales forecasting. Approaches such as hesitant fuzzy linguistic preference relation and ordinal consensus measure are used in our paper. Decision-makers compare alternatives over each criterion and the ranking of alternatives can be derived. Based on the ranking provided by the decision-makers, we introduce the ordinal consensus of the group. Then, a consensus reaching process is given to raise the degree of consensus.

Keywords Multiple criteria group decision-making · Fashion sales forecasting
Hesitant fuzzy linguistic preference relation · Ordinal consensus

1 Introduction

Fashion sales forecasting refers to the forecast of sales in the future based on a series of criteria. Due to the increasingly intensive competition, sales forecasting plays a more and more irreplaceable role in fashion enterprises [1]. Accurate sales predication can help enterprise leaders to develop reasonable sales strategy, which can further result in quick response to the market and low inventory levels [2]. However, fashion sales forecasting is a complex problem due to the uncertain and rapidly changing

M. Tang · H. Liao (✉)
Business School, Sichuan University, Chengdu 610064, China
e-mail: liaohuchang@163.com

M. Tang
e-mail: tangming0716@163.com

© Springer Nature Switzerland AG 2019
W. K. Wong (ed.), *Artificial Intelligence on Fashion and Textiles*,
Advances in Intelligent Systems and Computing 849,
https://doi.org/10.1007/978-3-319-99695-0_40

environment. A good sales forecasting need to consider a variety of factors and adopt the knowledge and experiences of a group of decision-makers (DMs). Multiple criteria group decision-making (MCGDM), which combines multiple criteria and group decision-making problems, is an important research content in the field of decision-making. DMs make judgments over each alternative with respect to different criteria according to their preferences. Then, the different individual preferences are aggregated into group preferences. In this way, the best alternative can be selected according to the group preferences. MCGDM is widely used in political, military, management, economics, and finance [3].

Artificial Intelligence (AI) is a smart machine for engineering and science. It is a technological science for developing the methods that are used for simulating and expanding human intelligence. AI consists of many branches such as artificial neural network, expert system, and fuzzy logic [4]. Fashion sales forecasting is a hot issue in fashion researching in recent years, and has become a critical factor that restricts the development of textile and fashion enterprises. Inaccurate forecasting may bring immeasurable consequences such as out of shock, high inventory and low utilization of resources. Due to the high uncertainty in the process of fashion sales and the short life cycle of fashion products, it is difficult to make a precise quantitative analysis for the fashion sales forecasting. The fuzzy linguistic logic that adopts experts' experience and knowledge would be helpful in fashion sales forecasting. The hesitant fuzzy linguistic term set (HFLTS) [5], which combines the fuzzy linguistic approach and context-free grammars, is a powerful tool to express people's qualitative information. Due to its advantage in expressing human's qualitative information, many HFLTS-related theories have been proposed [5–8]. A comprehensive overview can be found in Ref. [8]. Rodríguez et al. [9] first used the linguistic comparative expressions to enrich the elicitation of preference relations and defined the hesitant fuzzy linguistic preference relation (HFLPR).

Group decision-making (GDM) with HFLPRs has attracted many scholars attention [8]. In GDM, consensus process plays a critical role in obtaining the final decision [10]. Consensus can guarantee that the decision result be supported by all DMs. There are two main types of consensus measures for HFLPRs. One is based on the distance to the preference relations [11], the other is based on the distance between DMs [12]. However, we can also research the consensus measure based on the ranking of alternatives, which is called as ordinal consensus. This paper aims to investigate the ordinal consensus of HFLPRs and study the consensus reaching process. Then, the proposed consensus measure is used to address fashion sales forecasting. The main contributions of this paper can be summarized as follows:

(1) We investigate the ordinal consensus measure in the hesitant fuzzy linguistic environment. The individual ordinal consensus (IOC) and the group ordinal consensus (GOC) are introduced.

(2) We study the ordinal consensus reaching process. If the consensus level does not reach the predefined threshold, a feedback mechanism is introduced to assist the DMs to revise their preferences.

(3) We use our consensus reaching method to address the MCGDM problem for fashion sales forecasting.

The rest of this paper is structured as follows: Sect. 2 makes a brief review of the knowledge about the HFLTS and HFLPR. In Sect. 3, the ordinal consensus measure and the consensus reaching process are illustrated. Section 4 presents the calculation process to address the problem of fashion sales forecasting. Some concluding remarks are made in Sect. 5.

2 Preliminary

To address real-world situations in which DMs are hesitant among several linguistic terms to assess alternatives, Rodríguez et al. [5] initially presented the concept of HFLTS. Afterward, Liao et al. [13] defined the mathematical form of HFLTS as $H_S = \{\langle x, h_S(x)\rangle x \in X\}$, where $h_S(x) = \{s_{\phi_l}(x)|s_{\phi_l}(x) \in S,$ $\phi_l \in \{-\tau, \ldots, 0, \ldots, \tau\}, l = 1, 2, \cdots, L(x)\}$ is a set of possible degrees of the linguistic variable x to the linguistic term set $S = \{s_t|t \in \{-\tau, \ldots, 0, \ldots, \tau\}\}$. $h_S(x)$ is named as the hesitant fuzzy linguistic element (HFLE).

Rodríguez et al. [9] first used the HFLEs to represent the hesitant linguistic preferences of DMs over alternatives and defined the concepts of the HFLPR. Later, Zhu and Xu [14] represented the HFLPR as a matrix $H = (h_{ij})_{n \times n}$, where $h_{ij} = \{h_{ij}^l|l = 1, 2, \ldots, L_{h_{ij}}\}$ is a HFLE, denoting the hesitant degrees to which x_i is preferred to x_j. h_{ij} should satisfy $h_{ij}^{\sigma(l)} \oplus h_{ji}^{\sigma(l)} = s_0$, $h_{ii} = s_0$, $L_{h_{ij}} = L_{h_{ji}}$ and $h_{ij}^{\sigma(l)} < h_{ij}^{\sigma(l+1)}$, $h_{ji}^{\sigma(l+1)} < h_{ji}^{\sigma(l)}$.

In most cases, the number of linguistic terms in two HFLEs is different. To ensure the maneuverability of calculation between two HFLEs, Zhu and Xu [14] added some elements to the shorter HFLE which has a fewer number of elements, and then constructed the normalized HFLPR.

3 Ordinal Consensus Measure for HFLPRs

Generally speaking, the consensus is regarded as a complete agreement among all DMs concerning all alternatives. It can be defined as a mutual agreement among the DMs with all opinions being heard and addressed [15]. In most cases, full consensus may be difficult to achieve. Most of the existing consensus measures are based on the preference relations provided by DMs. There are several consensus measures for HFLPR [10, 11, 16]. The essence of these measures depends on the distance between HFLEs. We can reach a consensus level before the aggregating and ranking processes. However, we can also research the consensus based on the ranking of DMs, which is called as ordinal consensus. In the ordinal consensus ranking problem, s DMs rank n alternatives from the first to the nth. The object of ordinal consensus ranking is to find a consensus ranking for all alternatives [16]. Many approaches regarding ordinal consensus ranking have been introduced, which can be divided into two categories: one is the ad hoc *method,* the other is the *distance-based method* [17]. A comprehensive overview on the ordinal consensus ranking can be found in Ref. [17]. In this section, we defined a new ordinal consensus measure for HFLPRs.

Suppose that there are a set of alternatives $X = \{x_1, x_2, \ldots, x_n\}$. s DMs assess these alternatives and provide their preferences. Suppose that $r_i^{(k)}$ is the rank of the ith alternative provided by the kth DM. r_i^G is the rank of the ith alternative provided by the group. The weight vector of DMs is $\omega = (\omega_1, \omega_2, \ldots, \omega_s)^T$. Then, the individual ordinal consensus (IOC) level for each DM can be defined as

$$\text{IOC}^{(k)} = 1 - \frac{1}{\max_{n-2} +2(n-1)} \sum_{i=1}^{n} \left| r_i^G - r_i^{(k)} \right| \tag{1}$$

Then, we can define the group ordinal consensus (GOC) level for the group based on Eq. (1), which is as follows:

$$\text{GOC} = \sum_{k=1}^{s} \omega_s \left(1 - \frac{1}{\max_{n-2} +2(n-1)} \sum_{i=1}^{n} \left| r_i^G - r_i^{(k)} \right| \right) \tag{2}$$

where $\max_{n-2}$ is the maximum difference in the rankings when there are $n - 2$ alternatives. It is easy to find that $0 \leq \text{IOC}^{(k)}$, $\text{GOC} \leq 1$. The larger the $\text{IOC}^{(k)}$ is, the higher the level of consensus that DM_k has. The consensus level depends on the order of alternatives provided by the DMs. If $\text{IOC}^{(k)} = 1$, then DM_k has complete consensus with the group, i.e., his ranking of the alternatives is the same as the group. If $\text{IOC}^{(k)} = 0$, then DM_k has no consensus with the group. The maximum difference between two rankings can be represented as $\max_{n-2} +2(n-1)$.

In real-world situations, it is not realistic to require a DM to have full agreement with the group. Generally, we can set an acceptable threshold λ of IOC and GOC. If $\text{GOC} \geq \lambda$, then the group reaches an acceptable consensus. The value of λ is determined by the group of DMs. If the consensus degree does not reach the acceptable consensus threshold, then a consensus reaching process should be implemented.

Next, we introduce a feedback mechanism to reach the acceptable consensus degree. The feedback mechanism consists of two advice rules: identification rules and direction rules. The identification rules are used to identify DMs, alternatives, and pairs of alternatives that contribute less in reaching a high-level consensus.

(1) Identification rule for DMs. This identification rule is used to identify DM_k who does not reach the predefined threshold γ, which can be expressed as

$$\text{DM} = \{\text{DM}_k | \text{IOC}^{(k)} < \gamma, \ k = 1, 2, \ldots, s\}. \tag{3}$$

(2) Identification rule for alternatives. For DM_k, this identification rule is used to identify the alternatives that should be modified by DM_k. This identification rule is expressed as

$$AL = \{x_i | \max\{\left| r_i^G - r_i^{(k)} \right|\}, i = 1, 2, \ldots, n\} \tag{4}$$

(3) Identification rule for pairs of alternatives. For any alternative $x_i \in AL$, this rule identifies the compared alternative x_j and the position (i, j) that should be changed. The positions which should be changed are denoted as

$$PO_i = \{(i, j) | x_i \in AL \wedge \max \left| (r_i^{(k)} - r_j^{(k)}) - (r_i^G - r_j^G) \right| \} \tag{5}$$

The direction rules are used to provide suggestions for DMs to adjust their evaluations. Based on $(r_i^{(k)} - r_j^{(k)})$ and $(r_i^G - r_j^G)$, the direction rules are designed as follows:

(1) Direction rule 1. If $(r_i^{(k)} - r_j^{(k)}) < (r_i^G - r_j^G)$, then DM_k should increase the preference that alternative x_i to x_j.
(2) Direction rule 2. If $(r_i^{(k)} - r_j^{(k)}) < (r_i^G - r_j^G)$, then DM_k should decrease the preference that alternative x_i to x_j.

4 Fashion Sales Forecasting

In the fashion business, sales forecasting plays an important role for commercial enterprises [18]. Nowadays, the topic of how to develop accurate forecasting methods has been very hot. Up to now, many tools have been developed for fashion sales forecasting such as artificial neural network [19], extreme learning machine [20], and expert diagnostic system [21]. However, the existed methods are time consuming. In the fashion industry, real-time and prompt forecasting is very crucial because of the rapid renewal of fashion products. Therefore, in the process of sales forecasting, experts' experiences can be significantly helpful.

Suppose that a fashion enterprise has three kinds of fashion to sale. Three DMs evaluate the potential of sales according to three criteria: c_1 (product lifecycle), c_2 (year-on-year growth rate) and c_3 (consumer factor). The weight vector of these three criteria is $v = (0.4, 0.5, 0.1)^T$. The weight vector of the three DMs is $w = (1/3, 1/3, 1/3)^T$. The linguistic term set is $S = \{s_0 = $ very bad, $s_1 = $ bad, $s_2 = $ slightly bad, $s_3 = $ medium, $s_4 = $ slightly good, $s_5 = $ good, $s_6 = $ very good$\}$. Let $\lambda = 0.8$. The three DMs evaluate the alternatives and give the hesitant fuzzy linguistic preference relations as follows:

$$H_1^{C_1} = \begin{Bmatrix} s_3 & s_4 & s_5 \\ s_2 & s_3 & s_4, s_5 \\ s_1 & s_1, s_2 & s_3 \end{Bmatrix}, \quad H_1^{C_2} = \begin{Bmatrix} s_3 & s_4, s_5 & s_6 \\ s_1, s_2 & s_3 & s_3, s_4, s_5 \\ s_0 & s_1, s_2, s_3 & s_3 \end{Bmatrix}, \quad H_1^{C_3} = \begin{Bmatrix} s_3 & s_2, s_3 & s_3, s_4 \\ s_3, s_4 & s_3 & s_3 \\ s_2, s_3 & s_3 & s_3 \end{Bmatrix},$$

$$H_2^{C_1} = \begin{Bmatrix} s_3 & s_3, s_4 & s_4, s_5 \\ s_2, s_3 & s_3 & s_4 \\ s_1, s_2 & s_2 & s_3 \end{Bmatrix}, \quad H_2^{C_2} = \begin{Bmatrix} s_3 & s_4 & s_5 \\ s_2 & s_3 & s_3, s_4 \\ s_1 & s_2, s_3 & s_3 \end{Bmatrix}, \quad H_2^{C_3} = \begin{Bmatrix} s_3 & s_6 & s_4, s_5 \\ s_0 & s_3 & s_1, s_2 \\ s_1, s_2 & s_4, s_5 & s_3 \end{Bmatrix},$$

$$H_3^{C_1} = \begin{Bmatrix} s_3 & s_1, s_2 & s_0, s_1 \\ s_4, s_5 & s_3 & s_2 \\ s_5, s_6 & s_4 & s_3 \end{Bmatrix}, \quad H_3^{C_2} = \begin{Bmatrix} s_3 & s_2, s_3 & s_1 \\ s_3, s_4 & s_3 & s_2 \\ s_5 & s_4 & s_3 \end{Bmatrix}, \quad H_3^{C_3} = \begin{Bmatrix} s_3 & s_3, s_4 & s_5 \\ s_2, s_3 & s_3 & s_4 \\ s_1 & s_2 & s_3 \end{Bmatrix}.$$

To save space, the computing process is skipped. The ranking of these three DMs and the aggregating ranking are $DM_1: x_1 > x_2 > x_3$; $DM_2: x_1 > x_2 > x_3$; $DM_3: x_3 > x_2 > x_1$; $DM_G: x_1 > x_2 > x_3$. The individual consensus and the group consensus are $IOC^{(1)} = 1$, $IOC^{(2)} = 1$, $IOC^{(3)} = 0$, $GOC = 2/3$. Since $2/3 < 0.8$, the consensus degree does not reach the expected level. Next, we use the identification rules and direction rules to improve the consensus degree of the group.

(1) Identification rule for the DMs. Since $IOC^{(1)} < \lambda$, DM_3 should make some modifications.
(2) Identification rule for the alternatives. According to this rule, x_3 (or x_1) has the biggest difference between individual ranking and group ranking.
(3) Identification rule for the pairs of alternatives. Since $r_1^G - r_3^G = -2$, $r_1^{(3)} - r_2^{(3)} = 2$, the positions that should be modified are $PO_3 = \{(1, 3), (3, 1)\}$

Next, the directions rules are used to provide suggestions for DM_3.

Since $(r_1^G - r_3^G) < (r_1^{(3)} - r_2^{(3)})$, DM_3 should increase the preference that alternative x_1 to x_3.

Suppose that DM_3 provides the new preference degree between x_1 and x_3:

$$H_3^{C_1} = \begin{Bmatrix} s_3 & s_1, s_2 & s_3, s_4 \\ s_4, s_5 & s_3 & s_4 \\ s_2, s_3 & s_2 & s_3 \end{Bmatrix}, \quad H_3^{C_2} = \begin{Bmatrix} s_3 & s_2, s_3 & s_4 \\ s_3, s_4 & s_3 & s_3 \\ s_2 & s_3 & s_3 \end{Bmatrix}, \quad H_3^{C_3} = \begin{Bmatrix} s_3 & s_3, s_4 & s_5 \\ s_2, s_3 & s_3 & s_4 \\ s_1 & s_2 & s_3 \end{Bmatrix}$$

Then, we can obtain the new ranking of DM_3: $x_2 > x_1 > x_3$. The new ordinal consensus is $GOC = 5/6$. Because $5/6 > 0.8$, the consensus reaching process ends. The final ranking is $x_1 > x_2 > x_3$.

Moreover, we can also use Wu and Xu [11]'s consensus measures to solve this problem. In their method, the similarity degree between the two preferences was first defined. Then, the group hesitant fuzzy linguistic decision matrix was calculated by the hesitant fuzzy linguistic weight averaging operator. Based on this, they obtained the consensus degree of individual DM. To reach the predefined threshold, they developed two kinds of rules to help DMs to revise their preferences and make new judgments. The final result of Wu and Xu [11]'s method is the same as that of our proposed method. However, since Wu and Xu [13]'s method needs more times of iterations, our method is more feasible and efficient.

5 Conclusions

Fashion sales forecasting refers to the evaluation of fashion sales according to several criteria such as product lifecycle, year-on-year growth rate, and consumer factor. In this paper, we used an MCGDM method based on experts' experiences and knowledge to forecast fashion sales. Consensus can guarantee that the decision result be supported by all DMs. The ordinal consensus measure for HFLPRs was developed in this paper. In the case that the consensus level does not reach the acceptable level, a consensus reaching process was introduced to address this issue. Finally, we used our method to forecast fashion sales problem. Given that the DMs' knowledge and experiences played a key role in making a decision, real-time and prompt forecasting was the advantage of our method compared with other methods such as artificial neural network. Moreover, in our approach, the consensus reaching process can make the decision result more convincing and stable. Thus, it is a good choice for fashion sales forecasting.

Acknowledgements The work was supported by the National Natural Science Foundation of China (71501135, 71771156), the Scientific Research Foundation for Excellent Young Scholars at Sichuan University (No. 2016SCU04A23), the 2018 Key Project of the Key Research Institute of Humanities and Social Sciences in Sichuan Province (No. LYC18-02, No. DSWL18-2), and the Scientific Research Foundation for Excellent Young Scholars at Sichuan University (No. 2016SCU04A23).

References

1. Kuo, R.J., Xue, K.C.: A decision support system for sales forecasting through fuzzy neural networks with asymmetric fuzzy weights. Decis. Support Syst. **24**, 105–126 (1998). https://doi.org/10.1016/S0167-9236(98)00067-0
2. Choi, T.M., Chow, P.S.: Mean-variance analysis of quick response program. Int. J. Prod. Econ. **114**, 456–475 (2008). https://doi.org/10.1016/j.ijpe.2007.06.009
3. Liu, W.S., Liao, H.C.: A bibliometric analysis of fuzzy decision research during 1970–2015. Int. J. Fuzzy Syst. **19**, 1–14 (2017). https://doi.org/10.1007/s40815-016-0272-z
4. Adnan, M.R.H.M., Sarkheyli, A., Zain, A.M., Haron, H.: Fuzzy logic for modeling machining process: a review. Artif. Intelli. Rev. **43**, 345–379 (2015). https://doi.org/10.1007/s10462-012-9381-8
5. Rodríguez, R.M., Martínez, L., Herrera, F.: Hesitant fuzzy linguistic term sets for decision making. IEEE Trans. Fuzzy Syst. **20**, 109–119 (2012). https://doi.org/10.1109/TFUZZ.2011.2170076
6. Liao, H.C., Xu, Z.S., Zeng, X.J.: Hesitant fuzzy linguistic VIKOR method and its application in qualitative multiple criteria decision making. IEEE Trans. Fuzzy Syst. **23**, 1343–1355 (2015). https://doi.org/10.1109/TFUZZ.2014.2360556
7. Liao, H.C., Yang, L.Y., Xu, Z.S.: Two new approaches based on ELECTRE II to solve the multiple criteria decision making problems with hesitant fuzzy linguistic term sets. Appl. Soft Comput. **63**, 223–234 (2018). https://doi.org/10.1016/j.asoc.2017.11.049
8. Liao, H.C., Xu, Z.S., Enrique, H.V., Herrera, F.: Hesitant fuzzy linguistic term set and its application in decision making: a state-of-the art survey. Int. J. Fuzzy Syst. (2018). https://doi.org/10.1007/s40815-017-0432-9

9. Rodríguez, R.M., Martínez, L., Herrera, F.: A group decision making model dealing with comparative linguistic expressions based on hesitant fuzzy linguistic term sets. Inform. Sci. **241**, 28–42 (2013). https://doi.org/10.1016/j.ins.2013.04.006

10. Liao, H.C., Xu, Z.S., Zeng, X.J., Merigó, J.M.: Framework of group decision making with intuitionistic fuzzy preference information. IEEE Trans. Fuzzy Syst. **23**, 1211–1227 (2015). https://doi.org/10.1109/TFUZZ.2014.2348013

11. Wu, Z.B., Xu, J.P.: Managing consistency and consensus in group decision making with hesitant fuzzy linguistic preference relations. Omega **65**, 28–40 (2016). https://doi.org/10.1016/j.omega.2015.12.005

12. Wu, Z.B., Xu, J.P.: An interactive consensus reaching model for decision making under hesitation linguistic environment. J. Intell. Fuzzy Syst. **31**, 1635–1644 (2016). https://doi.org/10.3233/JIFS-151708

13. Liao, H.C., Xu, Z.S., Zeng, X.J., Merigó, J.M.: Qualitative decision making with correlation coefficients of hesitant fuzzy linguistic term sets. Knowle. Based Syst. **76**, 127–138 (2015). https://doi.org/10.1016/j.knosys.2014.12.009

14. Zhu, B., Xu, Z.S.: Consistency measures for hesitant fuzzy linguistic preference relations. IEEE Trans. Fuzzy Syst. **22**, 35–45 (2014). https://doi.org/10.1109/TFUZZ.2013.2245136

15. Saint, S., Lawson, J.R.: Rules for reaching consensus: a modern approach to decision making. Jossey-Bass, San Francisco (1994)

16. Xu, Y.J., Wang, H.M.: A group consensus decision support model for hesitant 2-tuple fuzzy linguistic preference relations with additive consistency. J. Intell. Fuzzy Syst. **33**, 41–54 (2017). https://doi.org/10.3233/JIFS-161029

17. Cook, W.D.: Distance-based and ad hoc consensus models in ordinal preference ranking. Eur. J. Oper. Res. **172**, 369–385 (2006). https://doi.org/10.1016/j.ejor.2005.03.048

18. Xia, M., Zhang, Y.C., Weng, L.G., Ye, X.L.: Fashion retailing forecasting based on extreme learning machine with adaptive metrics of inputs. Knowle. Based Syst. **36**, 253–259 (2012). https://doi.org/10.1016/j.knosys.2012.07.002

19. Chang, P.C., Wang, Y.W.: Fuzzy Delphi and back-propagation model for sales forecasting in PCB industry. Expert Syst. Appl. **30**, 715–726 (2006). https://doi.org/10.1016/j.eswa.2005.07.031

20. Sun, Z.L., Choi, T.M., Au, K.F., Yu, Y.: Sales forecasting using extreme learning machine with applications in fashion retailing. Decis. Support Syst. **46**, 411–419 (2008). https://doi.org/10.1016/j.dss.2008.07.009

21. Lin, C.T., Lee, I.F.: Artificial intelligence diagnosis algorithm for expanding a precision expert forecasting system. Expert Syst. Appl. **36**, 8385–8390 (2009). https://doi.org/10.1016/j.eswa.2008.10.057

Probabilistic Linguistic Linear Least Absolute Regression for Fashion Trend Forecasting

Lisheng Jiang, Huchang Liao◉ and Zhi Li

Abstract Fashion trend is an important aspect in costume designing given that the correct fashion trend prediction can help productions to occupy markets in short time. In the methods of forecast, fuzzy linear least absolute regression is a useful model. Meanwhile, most descriptions about the fashion trend are in nature words which are difficult to be used directly in present models. To deal with this problem, the probabilistic linguistic term set, a powerful tool in expressing and computing nature language, is introduced in this paper. First, operations on probabilistic linguistic term sets are modified to be more logical in the solution procedure of regression. Then a novel model which combines fuzzy linear least absolute regression and probabilistic linguistic term set is developed. Finally, an illustration about the forecast of clothing fashion trend is given to show the applicability of our method in costume designing evaluation.

Keywords Fashion trend forecasting · Fuzzy linear least absolute regression
Probabilistic linguistic term set · Fuzzy-in and fuzzy-out

1 Introduction

Probabilistic linguistic term set (PLTS) [1] can reflect more vague information in nature language and more probabilistic information compared with the crisp values and the hesitant fuzzy linguistic term set [2] respectively. Because of the precision and completeness in information representation, the PLTS has been investigated by many scholars [3] and has been applied in many practical problems, such as the hospital management [4] and the risk assessment [5].

L. Jiang · H. Liao (✉) · Z. Li
Business School, Sichuan University, Chengdu 610064, China
e-mail: liaohuchang@163.com

L. Jiang
e-mail: lsjiang96@163.com

© Springer Nature Switzerland AG 2019
W. K. Wong (ed.), *Artificial Intelligence on Fashion and Textiles,*
Advances in Intelligent Systems and Computing 849,
https://doi.org/10.1007/978-3-319-99695-0_41

Fuzzy regression analysis [6] is very useful in setting up a system structure of fuzzy logic. It can deal with the problems of the classical regression models which are with low sample size and with the vagueness which does not obey the random errors [7]. Fuzzy linear least absolute regression [8] is a method of fuzzy regression analysis which is preferred when the data are fat-tailed or outlier-produced. Besides, it is proved that the fuzzy linear least absolute regression can be changed into a linear programming [9] which is easy to work out.

Though there are many studies in the fuzzy regression analysis, little attentions are paid on the regression over linguistic terms. However, in practical, information cannot be transformed into precise data completely. In this situation, the linguistic terms are unavoidable. Due to this reason, in this paper, a novel method named the probabilistic linguistic linear least absolute regression (PLLLAR) is proposed. In this method, the input and output data are PLTSs while the regression parameters are crisp values. To combine the PLTSs with the fuzzy linear least absolute regression, basic operations on the PLTSs are modified since the present operations may cause model errors in determinant computation procedures. Then, the PLTSs are handled by the fuzzy linear least absolute regression successfully. After that, an illustration about the fashion trend predication is given to show the applicability of the proposed method. The contributions of this paper can be summarized as follows:

(1) Fundamental operations on PLTSs are altered. Based on these changes, the computation process of linear programming is accurate and the errors caused by the model structure are reduced.
(2) The procedure of the PLLLAR is developed. Furthermore, a linear programming is proposed based on this model.
(3) The forecast of the fashion trend which involves much qualitative information is solved. Hence, in a fashion period, the popular degree in the future is securable.

This paper is organized as follows: In Sect. 2, we review some elementary knowledge of PLTS and the development of fuzzy linear least absolute regression. Meanwhile, the fashion trend forecast is described in this section. Section 3 defines new operations of the PLTSs. The PLLLAR method is developed in Sect. 4. In Sect. 5, a numerical illustration concerning the fashion trend forecast is given to show the real-life application in aiding decision making of clothing designing. The paper ends with some concluding remarks in Sect. 6.

2 Preliminary

2.1 Probabilistic Linguistic Term Set

Let $S = \{s_t | t = -\tau, \ldots, -1, 0, 1, \ldots, \tau\}$ be a LTS. A PLTS is defined as [1] $L(p) = \{L^k(p^k) | L^k \in S, 0 \leq p^k \leq 1, k = 1, 2, \ldots, \#L(p), \sum_{k=1}^{\#L(p)} p^k \leq 1\}$, where $L^k(p^k)$ is the linguistic term L^k associated with its probability p^k. $\#L(p)$ is the amount of different linguistic terms in $L(p)$. The PLTS is originated from the hesitant fuzzy linguistic term set [3]. It gives a probability to each linguistic term in the PLTS. It is very similar to the Z-restriction [10], so that people can give their probability and possibility information.

2.2 Fuzzy Linear Least Absolute Regression

Fuzzy regression methods can be categorized into three groups. The first one is the possibilistic regression which was proposed by Tanaka et al. [6]. These methods minimize the vagueness of the model and change the optimization problem of estimation into a linear programming problem. The second group is the fuzzy least-square approaches, which are extensions of the classical least-square method. Both of these two types of approaches are impressible to outliners, and thus the fuzzy linear least absolute regression [8] was proposed.

In fuzzy regression, the inputs, outputs and parameters have different forms. Here we just concentrate on the category where the outputs and inputs are fuzzy information while the regression parameters are crisp values. Suppose that $(\tilde{x}_{ij}, \tilde{y}_i)\,(i = 1, 2, \ldots n;\ j = 1, 2, \ldots m)$ is the observed fuzzy information. $\widetilde{Y}_i\,(i = 1, 2, \ldots n)$ is the estimated fuzzy information. $A_j\,(j = 1, 2, \ldots m)$ are the regression parameters which are crisp numbers. Then, we have

$$\widetilde{Y}_i = A_0 + A_1\tilde{x}_{i1} + A_2\tilde{x}_{i2} + \ldots A_j\tilde{x}_{ij} \tag{1}$$

The aim of the fuzzy linear least absolute regression model is to minimize the absolute disparity between $\widetilde{Y}_i$ and $\tilde{y}_i$. Therefore, the objective function is:

$$\min|\widetilde{Y}_i - \tilde{y}_i| \tag{2}$$

2.3 Fashion Trend Forecasting

The change of culture, economic, and life style may influence the clothing fashion trend. The fashion trend may also have influence on the costume designing since some fashion elements are very important in a design. Hence, if the fashion trend is caught, the design would be successful. Then, why the fashion trend exists? Actually, the fashion trend is the appearance of the social development trend which involves politics, economics, and science technologies. In this sense, fashion trend is a whole feature of a specific period. In other words, it is a social phenomenon which is created by the consciousness of imitation to others.

Constrained and influenced by the religion, politics, economy, aesthetics and psychology, fashion trend prediction has some features [11]. First, it is scientific that most actions are regular and fashion trend itself is an imitation to others so the fashion trend is also regular. Based on these rules, the forecast is possible. The second feature is that there are some authoritative organizations which guide the fashion trend mostly. For instance, the Paris Fashion Week almost guides the trend in the following years. The last feature is called the incomplete truth. Here is an example that no matter what we predict the weather is, the weather will not change. This means if tomorrow is a rainy day, it will not change to a sunny day because of the forecast that tomorrow is a sunny day. In this condition, the weather is a complete truth. As for the fashion, it will change because of the predictions of some institutions such as the Paris Fashion Week. People's preferences will be changed because of these institutions' guidance. The reason for this phenomenon is that the body imitated by people is changed.

As the fashion trend is important and possible to be forecasted, two prediction characteristics are determined [12]. First, the fashion trend forecast is vague. The fashion trend has many themes and different groups of people have different themes. Because of this characteristic, the target of prediction may not be fixed, which causes the vagueness. The second is the artistry. Clothes are practical artwork and clothing design is practical artistic creation. The final result of the clothing designing is always shown by the designers' unique designs, obeying the forecast results. To make the designs artistic, in a way, the forecasting objects should be artistic.

Due to the importance of fashion trend prediction, many studies have been done on this issue. Using the machine learning and big data analysis, color trend [13], and skirt profile [14] have been analyzed, respectively.

3   New Operations of the PLTSs

Since the PLTS is a new linguistic information representation form, basic operations of the PLTSs need to be defined. Pang et al. [1] defined the addition of PLTSs. Then, Liao et al. [15] provided the subtraction of PLTSs.

Before we do the subtraction, we need to make sure that two PLTSs have the same lengths. Let $L_1(p_1) = \{L_1^{k_1}(p_1^{k_1})|L_1^{k_1} \in S, 1 \geq p_1^{k_1} \geq 0, k_1 = 1, 2, \ldots, \#L_1(p_1)\}$ and $L_2(p_2) = \{L_2^{k_2}(p_2^{k_2})|L_2^{k_2} \in S, 1 \geq p_2^{k_2} \geq 0, k_2 = 1, 2, \ldots, \#L_2(p_2)\}$ be two PLTSs with $\#L_1(p_1) = \#L_2(p_2)$. Then,

$$L_1(p_1) = \left\{ \begin{array}{l} L_1^{11}(\Delta), \ldots, L_1^{1i_1}(\Delta), \ldots, L_1^{k_1\left[\sum_{n_1=1}^{(k_1-1)} i_{n_1}+1\right]}(\Delta), \ldots, L_1^{k_1\sum_{n_1=1}^{k_1} i_{n_1}}(\Delta)| \\ \qquad k_1 = 1, 2, \ldots, \#L_1(p_1); i_{n_1} = \frac{p_1^{n_1}}{\Delta}, n_1 = 1, \ldots, k_1 \end{array} \right\}$$

$$L_2(p_2) = \left\{ \begin{array}{l} L_2^{11}(\Delta), \ldots, L_2^{1i_1}(\Delta), \ldots, L_2^{k_2\left[\sum_{n_2=1}^{(k_2-1)} i_{n_2}+1\right]}(\Delta), \ldots, L_2^{k_2\sum_{n_2=1}^{k_2} i_{n_2}}(\Delta)| \\ \qquad k_2 = 1, 2, \ldots, \#L_2(p_2); i_{n_2} = \frac{p_2^{n_2}}{\Delta}, n_2 = 1, \ldots, k_2 \end{array} \right\} \quad (3)$$

where $\Delta > 0$ is a sufficient small figure. In this condition, enough $s_\phi(\Delta)$ are added to the short one until both have equal length. Because $s_\phi(\Delta)$ means all linguistic terms are possible, it will not influence the total information.

The addition and subtraction of two PLTSs $L_1(p_1)$ and $L_2(p_2)$ can be defined as [15]

$$
L_1(p_1) \pm L_2(p_2) = \begin{cases} L_1^{11}(\Delta) \pm L_2^{11}(\Delta), \ldots, L_1^{1i_1}(\Delta) \pm L_2^{1i_1}(\Delta), \ldots, L_1^{k_1\left[\sum_{n_1=1}^{(k_1-1)} i_{n_1}+1\right]}(\Delta) \\ \pm L_2^{k_2\left[\sum_{n_2=1}^{(k_2-1)} i_{n_2}+1\right]}(\Delta), \ldots, L_1^{k_1\sum_{n_1=1}^{k_1} i_{n_1}}(\Delta) \pm L_2^{k_2\sum_{n_2=1}^{k_2} i_{n_2}}(\Delta) \end{cases}
$$

(4)

In the addition or subtraction operation, there are two problems that need to be solved. First, the length of the PLTSs must be the same, otherwise some terms cannot be calculated. The second is that only when the probability or the linguistic term is the same is the calculation effective. In Eq. (3), each probabilistic linguistic term is divided into several terms which have the same probability Δ. In this situation, all PLTSs are based on the same probability. In case that the split PLTSs do not have the same length, enough $s_\phi(\Delta)$ is added to the shorter PLTS. The computation shown in Eq. (4) is feasible since the length and the probability in PLTSs are the same.

Although the subtraction operation has some advantages especially when two PLTSs have different lengths, there is a drawback which cannot be ignored. For example, for two PLTSs $L_1 = \{s_0(0.2), s_1(0.3)\}$ and $L_2 = \{s_2(0.6)\}$, by Eq. (4), we obtain $L_1 + L_2 = \{s_2(0.3), s_3(0.3)\}$ and $L_1 - L_2 = \{s_{-2}(0.3), s_{-1}(0.3)\}$. Here, $(L_1 + L_2) + (L_1 - L_2) = 2 \times L_1 = \{s_2(0.3), s_3(0.3)\} + \{s_{-2}(0.3), s_{-1}(0.3)\} = \{s_0(0.3), s_2(0.3)\} \Rightarrow L_1 = \{s_0(0.3), s_1(0.3)\}$ but actually $L_1 = \{s_0(0.2), s_1(0.3)\} \neq \{s_0(0.3), s_1(0.3)\}$. Thus, the subtraction given by Liao et al. [15] is out of work in some situation. We need to modify the subtraction of the PLTSs.

Definition 1 Let $S = \{s_t | t \in \{-\tau, \ldots, -1, 0, 1, \ldots, \tau\}\}$ be a linguistic term set. Two PLTSs are $L_1(p_1) = \{s_t(p_1^t) | t \in \{-\tau, \ldots, -1, 0, 1, \ldots, \tau\}\}$ and $L_2(p_2) = \{s_t(p_2^t) | t \in \{-\tau, \ldots, -1, 0, 1, \ldots, \tau\}\}$. The subtraction is defined as

$$
L_1(p_1) \pm L_2(p_2) = \bigcup_{t=\{-\tau,\ldots,-1,0,1,\ldots,\tau\}} \{s_t(p_1^t \pm p_2^t)\},
$$

(5)

where p^t means the probability associated with the linguistic term s_t. In case the linguistic term s_t does not appear in the PLTS, we set $p^t=0$.

The difference between Eq. (4) and Eq. (5) is that they have different reference points. In any PLTS, there are two dimensions: linguistic term and probability, which cannot be converted into each other. Equation (4) is based on the probability, so only when the probability is the same can the linguistic terms be calculated. However, to keep the same length of PLTSs, some factitious operations is needed, which may lead to the wrong derivation as mentioned before. On the contrary, Eq. (5) is based on the linguistic term so the length of PLTSs are always the same since all the PLTSs are based on the same linguistic term set s. Hence, no factitious operations are involved.

Example 1 Give two PLTSs $L_1 = \{s_0(0.2), s_1(0.3)\}$ and $L_2 = \{s_2(0.6)\}$, by Eq. (5), we can get $L_1 + L_2 = \{s_0(0.2), s_1(0.3), s_2(0.6)\}$ and $L_1 - L_2 = \{s_0(0.2), s_1(0.3), s_2(-0.6)\}$ so $(L_1+L_2)+(L_1-L_2) = 2 \times L_1 = \{s_0(0.4), s_1(0.6)\} \Rightarrow L_1 = \{s_0(0.2), s_1(0.3)\}$. Here, in $L_1 + L_2$, the total probability is bigger than 1. This is because $0 \leq \sum p_1^t \leq 1, 0 \leq \sum p_2^t \leq 1 \Rightarrow 0 \leq \sum (p_1^t + p_2^t) \leq 2$. Meanwhile, in $L_1 - L_2$, the negative probability appears which means it is not possible with the probability 0.6.

4 Probabilistic Linguistic Linear Least Absolute Regression

Before introducing the PLLLAR method, the probabilistic linguistic regression matric should be given. For a event, let $\tilde{x}_{ij}(i = 1, 2, \ldots m; j = 0, 1, 2, \ldots n)$ be the ith decision maker's assessment under the jth criterion (C_j). $\tilde{y}_i(i = 1, 2, \ldots, m)$ is the overall judgment of the ith decision maker under all criterion. Here, in this paper, we just consider the fuzzy-in and fuzzy-out situation so $\tilde{x}_{ij}$ and $\tilde{y}_i$ are all in PLTSs. The standard form of the probabilistic linguistic regression matric is shown in Table 1.

Definition 2 Let $L(p) = \{s_t(p^t)|t \in \{-\tau, \ldots, -1, 0, 1, \ldots, \tau\}\}$ be a PLTS. The absolute value of the PLTS is defined as

$$|L(p)| = \{s_t(|p^t|)|t \in \{-\tau, \ldots, -1, 0, 1, \ldots, \tau\}\} \qquad (6)$$

where $|p^t|$ means the absolute value of s_t's probability.

Improving Eq. (1), we can get

$$\tilde{Y}_i = \{s_t(A_0 p_{i0}^t + A_1 p_{i1}^t + A_2 p_{i2}^t + \cdots + A_j p_{ij}^t)|t \in \{-\tau, \ldots, -1, 0, 1, \ldots, \tau\}\} \qquad (7)$$

where $A_j(j = 0, 1, 2, \ldots n)$ are crisp values and p_{ij}^t is the probability of the linguistic term s_t in the PLTS $\tilde{x}_{ij}$. Then, $\tilde{Y}_i - \tilde{y}_i = \bigcup_{t \in \{-\tau, \ldots, -1, 0, 1, \ldots, \tau\}} \{A_0 p_{i0}^t + A_1 p_{i1}^t + A_2 p_{i2}^t + \cdots + A_j p_{ij}^t - p_i^t\}$ where p_i^t is the probability of the linguistic term s_t of $\tilde{y}_i$. Finally, the objective function of the programming is

Table 1 Probabilistic linguistic regression matrix

No.	C_0	C_1	$\cdots$	C_n	Y
1	$\tilde{x}_{10}$	$\tilde{x}_{11}$	$\cdots$	$\tilde{x}_{1n}$	$\tilde{y}_1$
$\vdots$	$\vdots$	$\vdots$	$\ddots$	$\vdots$	$\vdots$
m	$\tilde{x}_{m0}$	$\tilde{x}_{m1}$	$\cdots$	$\tilde{x}_{mn}$	$\tilde{y}_m$

$$\min|\widetilde{Y}_i - \tilde{y}_i| = \sum_t |A_0 p_{i0}^t + A_1 p_{i1}^t + A_2 p_{i2}^t + \cdots + A_j p_{ij}^t - p_i^t| \qquad (8)$$

Furthermore, the programming can be transformed into a linear programming

$$\min = \sum (u_i^t + v_i^t)$$

s.t.

$$\begin{cases} A_0 p_{i0}^t + A_1 p_{i1}^t + A_2 p_{i2}^t + \cdots + A_j p_{ij}^t - p_i^t = u_i^t - v_i^t \\ u_i^t \geq 0 \\ v_i^t \geq 0 \end{cases} \qquad (9)$$

where u_i and v_i are PLTSs in which u_i^t and v_i^t are the probability of the linguistic term s_t, respectively. Model (10) can be solved by Lingo or Matlab.

5 Case Study on Fashion Trend Forecasting

In clothing designing, the prediction of fashion trend is very important because the design with suitable trend can be accepted quickly by the public and thus companies will have many benefits. However, the forecast of fashion trend is difficult because of two reasons. One is that fashion trend itself is a complex appearance, associated with the religion, politics, economy, aesthetics, and psychology. Hence, precise numbers cannot express the whole information, especially in religion and aesthetics, which may cause errors in prediction results. The second one is that fashion trend is not a complete truth. In this situation, it is not enough or exact if we just use the data in one or two years ago. Judgments of authoritative institutions are necessary and very important. Caused by these two reasons, the present methods with crisp numbers are not adequate. Since the PLLLAR method has advantages in dealing with qualitative regressions, the assessments of institutions can be involved in the PLLLRM to handle the problem of fashion trend forecasting.

To deal with this problem, the probabilistic linguistic regression matric should be given. Suppose that one institution is asked to give its assessments on nine designs under four criteria:

C_1 (shell fabric level): $\{s_{-2} =$ lowest; $s_{-1} =$ normal; $s_0 =$ medium; $s_1 =$ advanced; $s_2 =$ top$\}$;
C_2 (style): $\{ s_{-2} = H$ - type; $s_{-1} = X$ - type; $s_0 = A$ - type; $s_1 = T$ - type; $s_2 = O$ - type$\}$;
C_3 (price): $\{s_{-2} =$ cheap; $s_{-1} =$ normal; $s_0 = s_1 =$ high; $s_2 =$ expensive$\}$;
C_4 (popularity): $\{s_{-2} =$ low; $s_{-1} =$ normal; $s_0 =$ medium; $s_1 =$ quite populary; $s_2 =$ crazy$\}$.

The data is shown in Table 2.

Here, $\tilde{y}_i$ is the judgment in the column C_4 and $\tilde{x}_{i0} = \{s_{-2}(0.1), s_{-1}(0.1), s_0(0.1), s_1(0.1), s_2(0.1)\}$ is a role like the constant term.

Table 2 The assessments of nine designs by an authoritative institution

No.	C_1	C_2	C_3	C_4
1	$\{s_{-2}(0.7), s_{-1}(0.1)\}$	$\{s_{-2}(0.9), s_{-1}(0.1)\}$	$\{s_{-2}(0.5), s_{-1}(0.5)\}$	$\{s_{-1}(0.3), s_0(0.4), s_1(0.2)\}$
2	$\{s_{-2}(0.3), s_{-1}(0.5)\}$	$\{s_{-2}(0.8), s_{-1}(0.2)\}$	$\{s_{-2}(0.4), s_{-1}(0.3)\}$	$\{s_{-1}(0.1), s_0(0.5), s_1(0.2), s_2(0.2)\}$
3	$\{s_{-1}(0.6), s_0(0.2)\}$	$\{s_0(0.3), s_1(0.1)\}$	$\{s_{-1}(0.2), s_0(0.1)\}$	$\{s_0(0.3), s_1(0.2), s_2(0.2)\}$
4	$\{s_{-1}(0.3), s_0(0.3), s_1(0.3)\}$	$\{s_0(0.5), s_1(0.2)\}$	$\{s_0(0.3)\}$	$\{s_{-1}(0.3), s_0(0.1), s_1(0.2), s_2(0.3)\}$
5	$\{s_0(0.3), s_1(0.5)\}$	$\{s_0(0.1), s_1(0.6)\}$	$\{s_0(0.4)\}$	$\{s_0(0.1), s_1(0.3), s_2(0.5)\}$
6	$\{s_0(0.4), s_1(0.3), s_2(0.2)\}$	$\{s_1(0.3), s_2(0.1)\}$	$\{s_0(0.5), s_1(0.3)\}$	$\{s_0(0.2), s_1(0.2), s_2(0.4)\}$
7	$\{s_1(0.5), s_2(0.2)\}$	$\{s_1(0.4), s_2(0.5)\}$	$\{s_1(0.7)\}$	$\{s_0(0.1), s_1(0.1), s_2(0.7)\}$
8	$\{s_1(0.1), s_2(0.5)\}$	$\{s_2(0.3)\}$	$\{s_1(0.3), s_2(0.3)\}$	$\{s_2(0.9)\}$
9	$\{s_2(0.7)\}$	$\{s_2(0.9)\}$	$\{s_2(0.5)\}$	$\{s_1(0.2), s_2(0.7)\}$

By Eq. (9), the result is $\widetilde{Y}_i = 1 \times \tilde{x}_{i0} + 0.25 \times \tilde{x}_{i1} + 0.125 \times \tilde{x}_{i2} - 0.25 \times \tilde{x}_{i3}$. After this expression is worked out, any design's popularity can be predicted if the assessments of shell fabric, style, and price are available. In other aspect, the regression parameters can embody the main fashion elements. In this numerical case, the regression parameter of the shell fabric level is the highest, so it influences the popularity most. Due to this fact, in this fashion period, the fashion trend is mostly leaded by the shell fabric level. Moreover, the parameter of the price is -0.25 and this is because the high price may stop people's steps.

For one design, given that $\tilde{x}_1 = \{s_0(0.4)\}$, $\tilde{x}_2 = \{s_0(0.3), s_1(0.3)\}$ and $\tilde{x}_3 = \{s_2(0.8)\}$. Then, the popularity of this design is $\{s_{-2}(0.1), s_{-1}(0.1), s_0(0.575), s_1(0.475), s_2(-0.1)\}$. It is noticeable that the total probability is bigger than 1. This is because after the computation, the value range of the total probability is changed, which has been explained in Example 1. This result shows that this design is not possible to be very popular but it may be quite popular or, more likely, just catch the average level. Referring this result, the company can predict the benefits by the popularity and costume designers may improve their works by the regression parameters. If the regression parameter is high, designers can pay more attention to the associated criteria.

6 Conclusions

The PLLLAR is preferred when the regression data are qualitative and outlined. In this paper, we made the following contributions to the regression analysis:

(1) We modified the addition and subtraction operations of PLTSs to fit the calculation of the linear programming. Furthermore, the absolute value of PLTS was defined.
(2) We put forward the PLLLAR method and simplified it into a linear programming.
(3) We solved the forecast problem of fashion trend by using the new method. This methodology is possible to give opportunities to companies in occupying the coming markets.

Although the PLLLAR has forte in dealing with qualitative regression, we just considered the fuzzy-in and fuzzy-out situation in this paper. In the future, a complete method that includes qualitative and quantitate data can be studied.

Acknowledgements The work was supported by the National Natural Science Foundation of China (71501135, 71771156), and the Scientific Research Foundation for Excellent Young Scholars at Sichuan University (No. 2016SCU04A23).

References

1. Pang, Q., Xu, Z.S., Wang, H.: Probabilistic linguistic term sets in multi-attribute group decision making. Inform. Sciences. **369**, 128–143 (2016)
2. Rodríguez, R.M., Martínez, L., Herrera, F.: Hesitant fuzzy linguistic term sets for decision making. IEEE Trans. Fuzzy Syst. **20**(1), 109–119 (2012)
3. Liao, H.C., Xu, Z.S., Herrera-Viedma, E., Herrera, F.: Hesitant fuzzy linguistic term set and its application in decision making: a state-of-the-art survey. Int. J. Fuzzy Syst. (2017). https://doi.org/10.1007/s40815-017-0432-9
4. Liao, H.C., Jiang, L.S., Xu, Z.S., Xu, J.P., Herrera, F.: A probabilistic linguistic linear programming method in hesitant qualitative multiple criteria decision making. Inform. Sciences. **415–416**, 341–355 (2017)
5. Zhang, Y.X., Xu, Z.S., Wang, H., Liao, H.C.: Consistency-based risk assessment with probabilistic linguistic preference relation. Appl. Soft Comput. **49**, 817–833 (2016)
6. Tanaka, H., Uejima, S., Asai, K.: Linear regression analysis with fuzzy model. IEEE Trans. Syst. Man Cybern. **12**, 903–907 (1982)
7. Pourahmad, S., Ayatollahi, S.M.T., Taheri, S.M., Agahi, Z.H.: Fuzzy logistic regression based on the least squares approach with application in clinical studies. Comput. Math Appl. **62**, 3353–3365 (2011)
8. Zeng, W.Y., Feng, Q.L., Li, J.H.: Fuzzy least absolute linear regression. Appl. Soft Comput. **52**, 1009–1019 (2017)
9. Charnes, A., Cooper, W.W., Ferguson, R.: Optional estimation of executive compensation by linear programming. Manage. Sci. **2**, 138–151 (1995)
10. Zadeh, L.: A note on Z-numbers. Inform. Sci. **181**, 2923–2932 (2011)
11. Jiang, S.Y.: Analysis of the importance of fashion trends in clothing design. West. Leather. **67** (2017)
12. Han, J.Y.: A preliminary study on fashion trend prediction. The new World Forecast. **4**, 56–58 (1991)
13. Mello, P., Storari, S., Valli, B.: Application of machine learning techniques for the forecasting of fashion trends. Intelligenza Artificiale. (2008)
14. Jia, S., Zhu, S.G., Victor, K.: Based on the transformation and upgrading of garment industry in PingHu, a new prediction method of fashion trend is analyzed: an example of female dress profile analysis. Spec. Topic Discuss. **5**, 81–88 (2016)
15. Liao, H.C., Jiang, L.S., Benjamin L.: Probabilistic linguistic ELECTRE III based on new operations. IEEE Trans. Fuzzy Syst. Technique Report

Author Index

© Springer Nature Switzerland AG 2019
W. K. Wong (ed.), *Artificial Intelligence on Fashion and Textiles*,
Advances in Intelligent Systems and Computing 849,
https://doi.org/10.1007/978-3-319-99695-0

Printed in the United States
By Bookmasters